PHYLOGENIES IN ECOLOGY

PHYLOGENIES IN ECOLOGY

A GUIDE TO CONCEPTS AND METHODS

Marc W. Cadotte and T. Jonathan Davies

PRINCETON UNIVERSITY PRESS

PRINCETON AND OXFORD

Library of Congress Cataloging-in-Publication Data

Names: Cadotte, Marc William, 1975– , author. | Davies, T. Jonathan, 1973– , author.
Title: Phylogenies in ecology : a guide to concepts and methods /
Marc W. Cadotte and T. Jonathan Davies.
Description: Princeton : Princeton University Press, 2016. |
Includes bibliographical references and index.
Identifiers: LCCN 2016007387 | ISBN 9780691157689 (hardcover : alk. paper)
Subjects: LCSH: Phylogeny. | Ecology. | Evolution (Biology)
Classification: LCC QH367.5 .C33 2016 | DDC 576.8/8—dc23
LC record available at https://lccn.loc.gov/2016007387

British Library Cataloging-in-Publication Data is available

This book has been composed in Minion Pro

Printed on acid-free paper. ∞

Printed in the United States of America

1 3 5 7 9 10 8 6 4 2

*To all of our intellectual and biological ancestors and descendants;
we are part of something bigger because of our nonindependence.*

CONTENTS

◇◇◇◇◇◇◇◇◇◇◇◇◇◇◇◇◇◇

PREFACE

Over the past 15 years, the merging of phylogenetic trees with ecology has led to new questions and insights about how nature is structured around us. But perhaps more importantly, this new field of phylogenetic ecology—*ecophylogenetics*—represents a reunification of the fields of ecology and evolution. Long treated as separate disciplines, the plethora of observations, experiments, and theoretical developments in ecology and evolution reinforce the reality that there are a finite number of mechanisms that shape the distribution and diversity of life on Earth. These mechanisms result in ecological and evolutionary patterns of diversity, and ecophylogenetic approaches offer an opportunity to integrate processes and patterns operating at multiple temporal and spatial scales.

This is not to say that there haven't been missteps, blind alleys, and problematic assumptions during the development of the field. Rather, as with any nascent discipline, the field of ecophylogenetics has had its ups and downs; some researchers have rightfully pointed out that early hypotheses were often unrealistically simplistic and, mirroring the progress of community ecology, others have (rather prematurely) announced the death of the field. However, ecophylogenetic patterns hold more information than measures of species richness and abundance, upon which much of ecology rests. The existence of phylogenetically nonrandom species associations is widespread in nature, and this reality provides an exciting opportunity to understand how communities are structured and assembled.

This book is meant to provide those interested in phylogenetically nonrandom patterns in ecology with the conceptual underpinnings and tools to pursue their own research on the topic. We also carefully highlight problematic assumptions and areas that need further research. In providing these tools, we exclusively use the R statistical programing language. There are other programs and platforms one could use, but we chose R because of our familiarity with it and the fact that it is now the most widely used statistical program in ecology and evolution. Further, since R is open sourced, new functions are contributed to R packages as quickly as they are created, facilitating further advancement. The R package is also freely available. However, despite all these important benefits, there are potential problems. It may be that user-contributed functions have bugs, and the onus is upon the user to ensure that a particular function performs the task asked of it. It is also useful to remember that when writing R scripts, the successful execution of a function or a few lines of code does not guarantee a correct answer. Further, and quite frustratingly, interdependency among packages often means that substantial changes to one package renders others inoperable. It may very well be that shortly after this book is published some of our coded examples fail due to changes in functions, object types, or required arguments. We will do our best to keep code updated in our repository linked here (http://press.princeton.edu/titles/10775.html).

This book has benefited from the help, feedback, and encouragement from a number of people. Firstly, we would like to thank all those researchers who have unselfishly contributed R code and created packages so that others can do better science. In particular, we acknowledge the contribution of a few researchers who have been instrumental in developing many of the methods and libraries we use in the following chapters, including Steve Kembel, Matt Helmus, Liam Revell, Luke Harmon, and William Pearse. We apologize to the many others we have not named; the length of this list is an indication of the field's strength and depth. Science is truly a cooperative enterprise. We are thankful to those who gave us permission to use their data or R scripts in this book, including Elsa Cleland for the Jasper Ridge data, and Shai Meiri for sharing data on the mammal community at Yotvata. This book greatly benefitted from discussions with Steve Kembel and Pedro Peres-Neto. We wish to thank Jie Liu and Xingfeng Xi for catching errors—though those that remain are completely our fault. We are also extremely grateful to Matthew Leibold and an anonymous reviewer for supplying comments on a draft of the book. For their continuous encouragement and overlooking repeatedly missed deadlines, we are grateful to Alison Kalett and Betsy Blumenthal at Princeton University Press. We are extremely grateful to Sheila Ann Dean for carefully copyediting this book. Finally, Marc would like to thank Shirley Lo-Cadotte for all of her love, patience, and support, but mostly her patience. Three and half years was a long time to have a distracted partner. Jonathan gives thanks to LW.

Marc Cadotte
Toronto, Canada

Jonathan Davies
Montreal, Canada

July 2015

PHYLOGENIES IN ECOLOGY

CHAPTER 1

An Entangled Bank

Evolutionary Relationships and Ecological Patterns

One faces the future with one's past.
—*Pearl S. Buck*

All organisms are the product of their own history. What they eat, where they live, the diseases they carry, and how they compete are all shaped by the environments and biological interactions experienced by their ancestors. For much of its scientific history the focus of ecological study was largely restricted to explaining the interactions between species and their environments; but it paid scant attention to patterns of shared evolutionary histories, and the integration of evolutionary theory into ecology began only recently. When defining the scope of ecology in the inaugural issue of the journal *Ecology* in 1920, Barrington Moore, the forester and fourth president of the Ecological Society of America, said "all life is controlled by two great forces, heredity and environment, and ecology is the science dealing with the environment." He was advocating specialization in the emergence of ecology as a scientific discipline. Perhaps the early divide between ecology and evolution was really about the differences in scientific cultures, whereby evolutionists viewed ecology as field based, primitive, and unscientific, and ecologists viewed evolutionary biologists as pursuing overly simplistic laboratory experiments and explanations (Kohler 2002). Or perhaps early ecologists sought specialization as a necessary way to develop and test key hypotheses, to craft new tools, and to create robust explanations for patterns of nature. As independent areas of research, ecological and evolutionary studies have provided profound insights into the formation and function of biological systems. For example, ecologists have developed predictions about species coexistence based on the ratios of limiting resources (Tilman 1982) and several theories of macroecology explain large-scale diversity patterns (Blackburn and Gaston 2006). Despite, or even because of this early history, researchers today are working in an era of synthesis, where Moore's two great forces are now part of unified explanations in ecology (Ricklefs 2007).

Evolutionary history has long informed our view of the world, and its consideration can be traced back in the scientific literature at least as far as Charles Darwin's *On the Origin of Species* (Darwin 1859). More recently, we have been witness to a rapid development of phylogenetic methods allowing phylogenetically informed comparisons of traits among groups of species (Felsenstein 1985, Harvey and Pagel 1991). Critically, concomitant advances in technologies have provided us with the raw material required for these new approaches—molecular sequences and phylogenetic trees. Interest in using phylogenetic information to evaluate mechanisms of community assembly and coexistence (Webb 2000) and niche conservatism (Holt 1996, Wiens and Graham 2005) increased dramatically as phylogenetic information for entire regions, clades, and communities has become readily available (fig. 1.1).

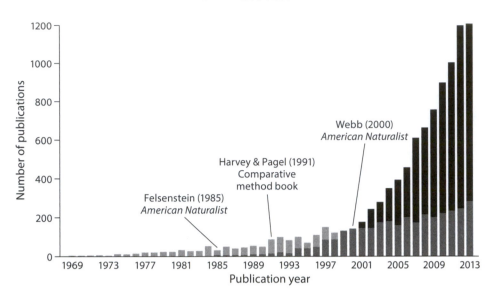

Figure 1.1. The number of publications indexed in Thompson Reuters ISI Web of Science that contain the keywords "phylogen'*'" and "community" (*dark bars*) from 1969–2013. The lighter bars transposed on top show the number of publications that contain the key word "competitive exclusion" for comparison. The dates of three classic publications that used phylogenetic information to test ecological hypotheses and thereby shaped the field are indicated.

Here we briefly review the history of the use of phylogenetics in ecology, starting with early attempts to classify the diversity of life and the development of evolutionary theory, through the rise of the comparative method, and finally to the emergence of ecological phylogenetics or *ecophylogenetics*.

1.1. SYSTEMATICS AND THE DIVERSITY OF LIFE

Even before evolutionary theory was widely accepted, early taxonomists, beginning with Carl Linnaeus, the father of modern taxonomy, have grouped species by similarity in their traits. Taxonomic ranks therefore represented a nested set of groupings—the Linnaean classification system (*Systema Naturae*)—from subspecies to the kingdoms of life. Linnaeus formally recognized only two kingdom's (Vegetabilia and Animalia) because the diversity of microscopic life was as yet largely unknown. A third kingdom was proposed to categorize the diversity of rocks and minerals, including fossils, but has fallen into disuse. We now recognize six separate kingdoms (Bacteria, Protozoa, Chromista, Plantae, Fungi, and Animalia), of which macroscopic organisms contribute just a small fraction—most of the life that surrounds us is invisible to us. A first step to categorizing the diversity of life is the placement of organisms within this taxonomic framework based upon their shared features.

The early classification presented in the 1735 *Systema Naturae* is now barely recognizable; some animal groups are still in use, including Aves (birds) and Mammalia (mammals), and although Linnaeus correctly placed bats (formerly grouped with birds) within Mammalia, whales were classified as fishes. In 1753, Linnaeus followed his *Systema Naturae* with publication of the *Species Plantarum*, in which he formally classified all the then known plants of Europe based predominantly on their reproductive traits. Plant taxonomy has gone through numerous iterations and currently the most widely accepted, traditional taxonomic treatment (i.e., based upon shared features) for flowering plants is that of Cronquist (Cronquist 1968). The revolution in molecular methods and the subsequent rapid expansion in the volume of gene sequence data then inspired the first major revision to the taxonomic system since Linnaeus. The Angiosperm Phylogeny Group, a loosely coordinated network of research institutions led by Mark Chase at the Royal Botanic Gardens, Kew, Peter Stevens from the Harvard University Herbaria, and Kåre Bremer from Uppsala University, proposed a new classification of flowering plants explicitly based on monophyletic groupings reflecting evolutionary relationships (APG 1998). The publication of this work inspired the following headline: "A Rose Is Still a Rose, but Everything Else in Botany Is Turned on its Head" from one national newspaper (*The Independent*, 23 November 1998). Indeed, a revised taxonomy based on DNA data revealed several surprising relationships. For example, despite apparent morphological and ecological similarities with water lilies (Nymphaeaceae), the sacred lotus (*Nelumbo*) was shown to be more closely related to the plane tree (*Platanus*) and the southern-hemisphere *Protea* (Proteaceae); the traits that had led previous taxonomists to group them together reflect a remarkable case of convergent evolution, also referred to in the systematics literature as homoplasy. Nonetheless, perhaps the most astonishing, but least remarked on, feature of the new APG classification system was its broad agreement with previous classifications—many taxonomic groupings were found to represent monophyletic groups, defined by a shared common ancestor. The traits used by taxonomists were informative for describing important evolutionary relationships. Such traits are referred to as synapomorphies—derived traits that are similar due to inheritance from a shared common ancestor. In modern terminology we might now describe these traits as being phylogenetically conserved.

1.2. THE ORIGINS

As the father of evolutionary biology, Charles Darwin presented a framework to explain both modern patterns of biological form and function and gradients of biological diversity. Darwin's general theory of evolution through natural selection informed his understanding of the geographical distribution of species, as well as the outcomes of species interactions. For example, he stated that a "larger number of the very common and much diffused or dominant species will be found on the side of larger genera" (Darwin 1859, 54). That is, community dominance likely relies on inherited traits linked to species diversification, thus bridging the gap between evolutionary process and ecological pattern. Further, he noted, "if a plot of bare ground be sown with one species of grass, and a similar plot be sown with several distinct genera of grasses, a greater number of plants and a greater weight of dry herbage can thus be raised" (Darwin 1859, 113). Here the central premise is that the more closely related two individuals are, the more likely they are to be functionally and ecologically similar; this idea laid the foundations for decades of research linking diversity and productivity.

Darwin's theory predicted a simple but fundamental relationship between evolutionary divergence and ecological distance between species. He stated this elegantly when concluding his thesis on the influence of relatedness and competition:

As species of the same genus have usually, though by no means invariably, some similarity in habits and constitution, and always in structure, the struggle will generally be more severe between species of the same genus when they come into competition with each other, than between species of distinct genera (Darwin 1859, 127).

Darwin thus recognized that a species' traits were important in determining its fit to the environment. Hence, closely related species sharing similar traits might be expected to be favored in similar environments; but also, competition would be greatest among species sharing similar ecologies, and patterns of evolutionary relationships serve as a representation of the ecological differences between species. The idea that ecological differences and evolutionary relatedness were critically important factors influencing species interactions became a central tenet in ecology by the mid-twentieth century. From his study of the birds in New Guinea, Ernst Mayr noted there was a general lack of co-occurrence between closely related species within habitats (Mayr 1942). Also, in a classic paper, Brown and Wilson (1956, 49) provided the slightly more formal observation on *closely related* species that have overlapping ranges. In the parts of the range where one species occurs alone, the populations of that species are similar to the other species and may even be very difficult to distinguish from it. In the area of overlap, where the two species occur together, the populations are more divergent.

Other early ecologists and biologists also saw the importance of relatedness in influencing ecological interactions and species distributions. For example, in discussing how competition structures animal communities, Charles Elton (1946, 64) commented:

Some (though not necessarily all) genera of the same consumer level that are capable of living in a particular habitat at all can coexist permanently on an area; whereas it is unusual, in the communities analyzed, for species of the same genus to coexist there.

Biogeographers seemed to readily embrace the importance of species relatedness to explain patterns of biodiversity. In contrast, the emergence of community ecology during this time rarely considered species' evolutionary relationships directly (but see Simberloff 1970, Jarvinen 1982). That is not to say that early ecologists were not concerned about similarities; to the contrary, early conceptual and theoretical approaches placed a great deal of emphasis on how species similarities and differences influenced coexistence (Grinnell 1917, 1925; Gause 1934; Macarthur 1958). For example, Robert MacArthur, building on the concepts of Grinnell, Gause, and Volterra, stated, "if the species requirements [of co-occurring species] are sufficiently similar . . . only one will be able to persist, so that the existence of one species may even control the presence or absence of another" (MacArthur 1958, 599). From these observations, evolutionary theory would suggest that the phylogenetic pattern of species coexistence and co-occurrence should be one of overdispersion, with communities of coexisting species comprising distant relatives. Separate from this early period in ecology, and coinciding with the emergence of biogeography and community ecology, a theory was developing on why some communities might be comprised of species that are more similar than expected by chance. This contrast perhaps reflected the more general dissociation between ecology and evolution during that time (Kohler 2002).

1.2.1. Neutral Dynamics and the Lack of Species Differences

While the importance of species differentiation led ecologists to predict that less-closely related species should be more able to coexist, the importance of ecological differences itself has been repeatedly questioned. At the same time as the rise and solidification of theories of coexistence based on niche differentiation, ecologists were observing community assemblages that were comprised of ecologically similar species (Ross 1957, Udvardy 1959) and started to explore the possibility that ecological interactions could be "neutral" (Sakai 1965). In 2001, Stephen Hubbell published *The Unified Neutral Theory of Biodiversity and Biogeography*, and introduced the neutral theory of biodiversity into mainstream ecology. Hubbell demonstrated that ecological communities could plausibly be the result of processes that rely little on interspecific differences, thus challenging the Hutchinsonian niche paradigm that had dominated community ecology over the preceding decades. The explicit assumption of neutral theory is that species have identical fitness responses to local environmental conditions. Under neutral processes one would therefore predict community assembly to be largely independent of phylogenetic history. However, neutral communities might still demonstrate strong phylogenetic structure. For example, phylogenetic structure can emerge through patterns of local dispersal. In addition, structure might reflect regional evolutionary dynamics; under a "point mutation" mode of speciation, where incipient species arise from a single individual, phylogenetic trees tend to be highly asymmetrical, dominated by one or a few species-rich clades (Mooers et al. 2007). In contrast, a fission mode of speciation, in which incipient species abundances may be large, can produce unusually symmetrical trees (Davies et al. 2011). Because tree symmetry also influences the evolutionary distances between taxa (Schweiger et al. 2008), depending upon the mode of speciation we might predict communities comprised of more or less related species, even when structured entirely by neutral processes (i.e., neither environmental filtering nor competition are important determinants of species coexistence).

1.2.2. Phylogenetic Patterns and Niche/Neutral Community Assembly

As focus shifted to consider multiple processes structuring species assemblages, the importance of spatial scale was also to become more apparent. The assembly of a local community is the result of the tension among local interactions, which are often negative, but sometimes neutral, and possibly deterministic (these interactions promote divergence and species ecological differences), as well as larger-scale environmental, dispersal constraints, or stochastic events (Berlow 1997, Belyea and Lancaster 1999, Levine 2000, Lovette and Hochachka 2006). At these larger scales, environment or dispersal limitation filters a larger pool of potential colonists into a smaller subset that includes species with the appropriate suite of traits for establishment (Keddy 1992, Weiher and Keddy 1995a), thereby promoting species similarities within communities. For this reason, larger regional species pools consisting of a broad array of functional traits should produce species assemblages with better matches between traits and local conditions compared to regions with small regional species pools (Questad and Foster 2008), perhaps shifting the balance between local versus large-scale processes.

The niche-based model of community assembly depends on phenotypic variability where niche differentiation maximizes differences among species, whereas neutral dynamics are the product of species similarities (MacArthur 1958, Schoener 1968, Fox and Brown 1993, Chesson 2000, Hubbell 2001, Chave 2004, Tilman 2004, Clark 2009). Recent work has shown that both evolutionary and ecological processes can give rise to assemblages built from processes simultaneously acting on species similarities and differences (Gravel et al. 2006, Holt 2006, Scheffer and van Nes 2006, Adler et al. 2007, Cadotte 2007, Fukami et al. 2007). Moreover, advances in coexistence theory suggest that species co-occurrence might reflect a balance between differences in competitive abilities and niche differences (Chesson 2000, Adler et al. 2007), which may themselves have opposing effects on relatedness patterns (Mayfield and Levine 2010). In addition, niche and fitness differences may evolve at different rates, and singular predictions about phylogenetic relatedness and the strength of competition are therefore unlikely to be correct (Gerhold et al. 2015). For example, species with high niche overlap may also be better competitors relative to other species because they share a common trait that gives them a competitive advantage (e.g., tall plants); consequently, communities structured by competition can contain similar species, which some may see as evidence for environmental filtering (Mayfield and Levine 2010).

While viewing community assembly as a product of species similarities and differences may be conceptually appealing, quantifying these differences in traits or ecological niches in diverse communities has proven difficult. When we measure a limited set of traits, we run the risk of missing important traits or over-inflating the role of measured traits (Wright et al. 2006, Mokany et al. 2008, Schamp et al. 2008, Cadotte et al. 2013, Kraft et al. 2015). Given our limited ability to adequately identify and measure species differences across diverse assemblages, ecologists are increasingly employing the evolutionary history of species (represented by their phylogenies) as a surrogate measure of the overall phenotypic similarities and differences among members of a community, or at least as a starting point to identify the most important phenotypic differences for community assembly (Webb 2000; Webb et al. 2002; Cavender-Bares and Wilczek 2003; Cavender-Bares et al. 2006; Lovette and Hochachka 2006; Helmus, Savage et al. 2007; Cavender-Bares et al. 2009). Phylogenetic approaches provide two principal advantages. First, phylogenetic patterns of community assembly can lead investigators to key traits or niche differences important for coexistence. Second, and more importantly, phylogenies may better capture the multitude of ecological requirements, tolerances, interactions and proclivities that define a species niche; they therefore provide a powerful tool for interpreting community assemblage, which is likely better than any single-trait studies. That is, the phylogeny of a regional pool of species represents the product of the phenotypic and niche variation among species.

1.3. "CORRECTING" ECOLOGICAL COMPARISONS

Despite increasing awareness of the importance of phylogenetic history in informing patterns of species distributions and co-occurrences, the first rigorous use of modern phylogenetic information in ecology was in a method to account for species covariance in multispecies comparisons. A long tradition in ecological and evolutionary research has been to determine how traits correlate with one another among multiple species. For example, there may be interest in relating brain size to body mass, egg clutch size to fledging time, or plant invasiveness to seed size. However, Felsenstein identifies a potentially serious problem when analyzing such data, arising "from the fact that species are part of a hierarchically structured phylogeny, and thus cannot be regarded for statistical purposes as if drawn independently from the same distribution" (Felsenstein 1985, 1; see also Harvey and Pagel 1991). That is, invasive plants may tend to have smaller seed sizes because, say, a large family has a preponderance to be invasive while also having small seeds without a mechanistic link between the two. Thus, invasiveness and small seeds are correlated because of shared ancestry, and in fact, any autapomorphy for the clade might correlate just as strongly with invasiveness; hence correlation does not imply causality. Since Felsenstein first articulated this problem in the statistical framework of phylogenetically independent contrasts, a number of solutions have been presented that account for phylogenetic covariance among species when measuring correlations among phenotypic traits (Felsenstein 1985, Harvey and Pagel 1991, Hansen 1997, Garland et al. 1999, Freckleton 2009, Nuismer and Harmon 2015).

1.3.1. Questioning the Relevance of Phylogenetic Corrections for Ecology

While the logic for the accounting of phylogenetic relationships in comparative analysis is persuasive, its relevance was called into question by some ecologists (Westoby et al. 1995, Ricklefs and Starck 1996). These researchers openly questioned the need for a "phylogenetic correction" of ecological analyses, for example when interpreting the importance of functional traits in different habitats. The concern was that phylogenetic corrections might erroneously weaken statistical relationships. In our above example, perhaps seed size does influence invasiveness, and applying the correction may obfuscate this relationship. The more robust approach would be to design experiments based upon the observed correlations. Philosophical considerations about the ability to infer the ecological value of traits aside, we cannot ignore the fact that failure to take phylogenetic relationships into account when examining trait correlations will lead to violations of the basic assumptions of most statistical tests (Ackerly and Donoghue 1995; Harvey et al. 1995b, 1995a; Ricklefs and Starck 1996). Rather than attempting to "correct" ecological analyses for phylogenetic nonindependence, modern comparative methods provide a statistical framework that allows phylogenetic information to be included within analyses of trait correlations (Ricklefs 1996, Ricklefs and Starck 1996, Martins and Hansen 1997, Abouheif 1999, Freckleton 2000).

1.4. THE EMERGENCE OF ECOPHYLOGENETICS

Perhaps the first published use of the term "ecophylogenetic" was by Armbruster (1992) in reference to the merging of phylogenetic information and ecological data so as to "infer the evolutionary history of ecological relationships." This early usage focused on the evolutionary origins of species' interactions. More recently the term ecophylogenetics has taken on a more specific definition, and is used most frequently to refer to phylogenetic community patterns and associated methods for assessing both potential assembly mechanisms and the key evolutionary events (e.g., trait evolution, species radiations, etc.) that influence the assembly of ecological communities (Cavender-Bares et al. 2009).

Campbell Webb's seminal paper in 2000 (Webb 2000) and the subsequent publication in 2002 of his influential review in the *Annual Review of Ecology and Systematics* (Webb et al. 2002) addressed whether communities contained species that were more or less closely related than expected by chance, and defined the modern field of ecophylogenetics. In 2000, Webb used a phylogeny of tree species found in a region of Borneo and generated null expectations for community phylogenetic patterns by creating random communities of co-occurring species. By comparing real community data to the null expectations, he showed that plots tended to contain species that were more closely related then expected by chance, inferring shared ecological preferences among close relatives. Previous work had approached similar questions using taxonomic information (Clarke and Warwick 1998, Warwick and Clarke 1998), or measured phylogenetic diversity as an alternative way to understand biological diversity (Faith 1992a, 1994; Crozier et al. 2005). However, Webb's framework, linking trait evolution with community assembly processes to predict patterns of phylogenetic relatedness in communities, along with the increasing availability of phylogenies for entire communities, led to a rapid proliferation of studies using phylogenetic information as a lens through which to view the distribution of biological diversity and interpret the structure of ecological communities (fig. 1.1).

Within the succeeding few years, a number of other key events helped shape the field and accelerate its rapid expansion. In 2006 the highly influential journal *Ecology* published a special issue entitled "Integrating Phylogenies into Community Ecology" (Webb et al. 2006). This showcased 14 examples of ecophylogenetic studies in various systems around the world, from bacteria to tropical forests, and bird assemblages to frog parasites. Over the past few years a growing a number of researchers and working groups supported by diverse institutes, including the Long Term Ecological Research (LTER), National Center for Ecological Analysis and Synthesis (NCEAS), and the German Center for Integrative Biodiversity (iDiv), have worked toward integrating phylogeny and community ecology. Ecophylogenetics is now an established paradigm under which ecologists design studies and interpret ecological patterns. Yet despite this recognition, there are numerous limitations and critical assumptions that require research priority.

1.5. THE GOAL OF THIS BOOK

This book is meant to be a synopsis of the major concepts and methods available for ecophylogenetic analyses. We assume that the ultimate goal is to use evolutionary history to gain a deeper understanding of extant ecological patterns. Put another way, we assume that the user wishes to test ecological hypotheses (as opposed to strictly evolutionary hypotheses) ranging from community assembly and dynamics, to ecosystem function, to macroecological patterns. While highlighting the use of phylogenies in ecology, we also carefully describe important assumptions and limitations.

We point to how ecophylogenetic analyses can be applied to understanding and predicting the effects of anthropogenic environmental change. For example, distributions and community composition are responding to species invasions, climate change, eutrophication, and disturbance, and these responses are dictated in large part by the species' evolutionary history. As we move into a world full of novel habitats, species are armed only with the information contained within their genomes—information derived from evolution in past environments.

We see this book as a potential resource for students and working scientists interested in learning about and applying phylogenetic approaches to analyzing ecological data. We assume that the reader is broadly familiar with modern ecology and its core concepts (e.g., competition, predation, dispersal, etc.). Further, we hope that this book will aid researchers in developing and refining their questions, and make clear the data and analyses necessary to answer these questions.

Broadly speaking, this book has three main goals. The first is as an introduction to the utility of phylogenies for understanding ecological patterns. Specifically, we explore topics such as how different assembly mechanisms generate communities, how large-scale diversity biodiversity gradients are generated and maintained, and how to analyze the distribution of ecologically relevant traits. Our second goal is to provide an instructional manual for the use of statistical methods for ecophylogenetic analysis. We will discuss the logic and assumptions of various tests, as well as constructing and interpreting different null modeling approaches. Our third goal is a practical one. We will detail the use of the tests in the R statistical programming language (www.R-project.org). All the examples used in this book will be fully implemented in R, and both the code and data are available on a companion website (http://press.princeton.edu/titles/10775.html). We assume that the user has a basic understanding of R, but we explain all the R code used in the examples so that even an R neophyte can implement the analyses.

1.5.1. Data Sets

Throughout this book we use several sample data sets to showcase methods and analyses. These data sets are available on the companion website. A brief note about the data sets:

1. Jasper Ridge: These data are from plant communities in 30 plots of 1×1 m^2 located at Stanford University's Jasper Ridge Biological Preserve in central California. For details of the data and the molecular phylogeny, see Cadotte et al. (2010b).
2. Phylogenetic supertree for mammals: This phylogenetic tree from Fritz et al. (2009) is a modification of the phylogenetic supertree originally published by Bininda-Emonds et al. (2007) with taxonomic updates. While a number of more recent mammal phylogenies have been published, this tree remains the only with almost complete taxonomic sampling, and is thus a valuable resource for comparative biologists. Three dated topologies are available; we use the "bestDates" version.
3. Global mammal distribution data: These data are derived from publicly available range map data made available through the IUCN (International Union for Conservation of Nature). Species ranges are depicted as polygons viewable in most common GIS software, and can be downloaded here: http://www.iucnredlist.org/technical-documents/spatial-data.
4. Mammal trait data: Tooth size data for Carnivora were collected by S. Meiri and were originally published in Davies et al. (2007). Body size data for the species community at Yotvata, Israel, were also collected by S. Meiri and were published in Davies, Cooper, et al. (2012).

CHAPTER 2

Building and Using Phylogenies

History is the science of what never happens twice.
—*Paul Valéry*

The fundamental premise of this book is that phylogenies provide a new framework for analyzing ecological data; this assumes that ecologists have phylogenies available to them and that they understand what the phylogeny represents. A phylogenetic tree is simply a representation of species interrelatedness (fig. 2.1) and it conveys information about which taxa share recent common ancestors, which evolutionary groups (clades) species belong to, and the distances (time, genetic, or character differences) separating species. In order to interpret the information within phylogenetic trees and know how this information can be used in ecological analyses, it helps to have a basic grounding in phylogenetic terminology. Figure 2.1 presents a sample phylogeny with various elements and terminology defined; these will appear frequently throughout this book.

While it is true that molecular data and phylogenetic trees are available like never before, it is not likely that a researcher will find a phylogeny that includes all the species on their species list. Researchers will often be required to build a phylogeny for their set of species. Depending on the information available, we may need to sequence tissue samples directly, search for and download sequences from online databases, use or patch together published trees, add species to a well-established backbone tree for higher-level taxa (e.g., families), or employ some combination of methods.

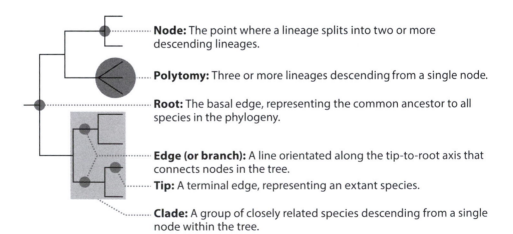

Node: The point where a lineage splits into two or more descending lineages.

Polytomy: Three or more lineages descending from a single node.

Root: The basal edge, representing the common ancestor to all species in the phylogeny.

Edge (or branch): A line orientated along the tip-to-root axis that connects nodes in the tree.

Tip: A terminal edge, representing an extant species.

Clade: A group of closely related species descending from a single node within the tree.

Figure 2.1. The anatomy of a phylogeny.

Once a phylogeny has been obtained, we still need to make decisions about whether and how to scale internal branch lengths, whether the tree should be measured in time or nucleotide substitutions, or if different clades have equivalent rates of evolutionary change. One aspect that should remain in the back of our minds is how these types of decisions might influence our ability to analyze and make inferences about the phenomena we may be interested in exploring.

2.1. HANDLING PHYLOGENIES IN R

Almost all of the various metrics, analyses and graphical representations that we explore in this book originate with the R phylogenetic object (referred to as a *phylo* object type). It is truly remarkable that this relatively simple object type can form the basis for the plethora of hypotheses and statistical methods in ecological and evolutionary analyses.

There are a number of phylogenetic packages available in R and it is not possible to review them all here because many do very specific things that we do not cover in this book, and many will likely change or be rendered obsolete over time. Instead, please refer to the CRAN task view for phylogenetic packages (http://cran.r-project.org/web/views/Phyloge-netics.html) maintained by Brian O'Meara. We will briefly discuss the two packages that we will use throughout this book. They are *ape* and *picante*. We will use other packages in various chapters, which we will briefly introduce as we use them.

2.1.1. Ape

By far the most frequently used package for phylogenetic analyses is *ape* (Analysis of Phylogenetics and Evolution; Paradis et al. 2004). *Ape* is a powerful, flexible package with many functions that allow you to import, visualize, manipulate, and distill information from phylogenetic trees. The package was created in 2002 by Emmanuel Paradis and colleagues and has grown over time in terms of the number of contributors and the number of functions it contains. The first version of *ape* contained less than 30 functions and the latest version at the time of writing (v. 3.3) contains over 250. It serves as the basis for handling phylogenies for a number of other packages, including *picante*, *geiger*, and *pez* (see section 2.1.3).

The key to *ape*'s widespread use is the ease with which it imports and handles phyloge-netic trees. To install and call this package, you can use the following commands:

```
install.packages("ape")
library(ape)
```

To see all the functions contained within *ape*, we can call the help function:

```
help(package=ape)
```

This opens an interface listing all functions and links to more detailed descriptions of those functions (fig. 2.2).

There are two functions to read in common tree formats; these are `read.nexus`, for nexus formatted trees, and `read.tree`, which reads in the common Newick format. All help files contain nine basic elements. The first three include (1) *description*, which is a sentence or two about what the function does; (2) *usage*, which shows how to use the function

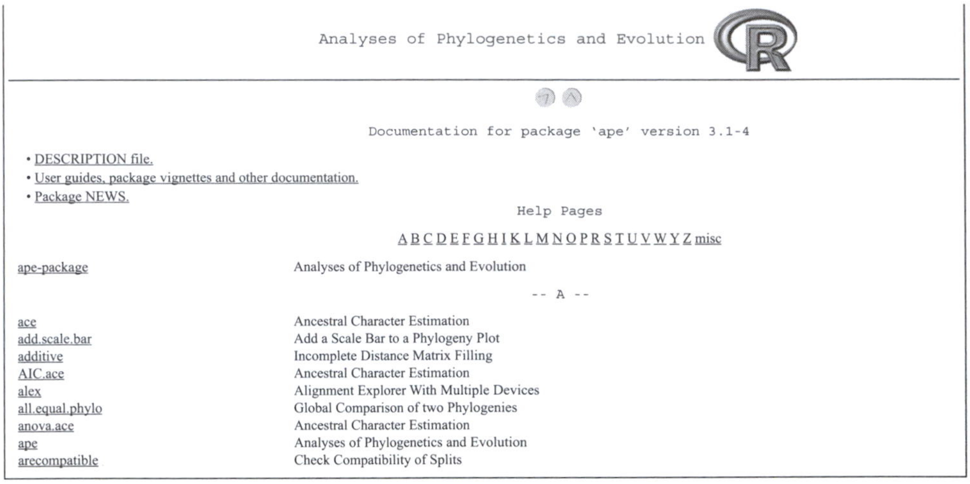

Analyses of Phylogenetics and Evolution

Documentation for package 'ape' version 3.1-4

- DESCRIPTION file.
- User guides, package vignettes and other documentation.
- Package NEWS.

Help Pages

A B C D E F G H I K L M N O P R S T U V W Y Z misc

ape-package   Analyses of Phylogenetics and Evolution

-- A --

ace   Ancestral Character Estimation
add.scale.bar   Add a Scale Bar to a Phylogeny Plot
additive   Incomplete Distance Matrix Filling
AIC.ace   Ancestral Character Estimation
alex   Alignment Explorer With Multiple Devices
all.equal.phylo   Global Comparison of two Phylogenies
anova.ace   Ancestral Character Estimation
ape   Analyses of Phylogenetics and Evolution
arecompatible   Check Compatibility of Splits

Figure 2.2. Interface for *ape* documentation. Functions are the links down the left side of the page.

in the command line interface and includes the various arguments contained within the function; and (3) *arguments*, which specifies what input arguments are required to run the function. Arguments either require input, such as `file = " "` in `read.tree`, or else have defaults that only need to be specified when we wish to specify an alternative assumption or value. The next two elements are (4) *details*, which gives more information about usage; and (5) *value*, which tells us what is created by running a function. In our case, `read.tree` returns an R object with a class definition "phylo" that is comprised of a number of components. We will go through these in detail below. The remaining elements are (6) *authors*, which tells us who wrote it; (7) *references*, which points us to pertinent literature or resources; (8) *see also*, which lists other related functions; and (9) *examples*, which gives a working example that we can copy and paste into R. All R function help files will have a working example.

The *phylo* object is how *ape* represents a phylogenetic tree, and it places the information contained in phylogenetic trees into six distinct components. First, let us load a tree. We will use the *ape* `read.tree` owl phylogeny example. You can either run the code in the `read.tree` help file, or load the same tree that we have supplied on the website:

```
tree.owls<-read.tree(file="owl_tree.phy")
```

If we type the object name "tree.owls" into the R interface, it returns summary information on the tree:

```
Phylogenetic tree with 4 tips and 3 internal nodes.

Tip labels:
[1] "Strix_aluco"  "Asio_otus"  "Athene_noctua"  "Tyto_alba"

Node labels:
[1] "5" "6" "7"

Rooted; includes branch lengths.
```

We can see that there are four tips or species, and R returns the names of the first four (in this case all) tip names. This summary also tells us that the tree is rooted and has branch lengths. The class of "tree.owls" is *phylo*, and this can be seen with:

```
class(tree.owls)

[1] "phylo"
```

The *phylo* object is a list with the aforementioned six components, which can be seen with:

```
attributes(tree.owls)

$names
[1] "edge" "Nnode" "tip.label" "edge.length"

$class
[1] "phylo"

$order
[1] "cladewise"
```

We will focus on the four *names* elements. They are easy to call using the $ symbol,

```
tree.owls$edge

      [,1]   [,2]
[1,]    5      6
[2,]    6      7
[3,]    7      1
[4,]    7      2
[5,]    6      3
[6,]    5      4
```

Edge is a matrix of node labels that are numbered 1 to *n*, where *n* is the total number of tips and internal edges, and the rows are the edges. In the summary of "tree.owls" we saw that there were four tips and three internal nodes, and 4 + 3 = 7. In this matrix the first column is the ancestor node and the second column is the descendant node. If there are *t* tips, then 1 to *t* are the tips. In our case, this is 1 to 4, and tips can only appear in the descendent (second) column since tips cannot be ancestors in a standard phylogenetic tree. The value *t* + 1, in our case 5, is the root or deepest node. In our tree, 5 leads to 6 and 4 (which is a tip), and 6 leads to 7 and 3 (which again is a tip), and so on. Figure 2.3 shows this tree with the nodes numbered.

Here is the code used to create figure 2.3. We added the node labels to the original owl tree file. For both files we added a scale bar and labeled it "MY" (millions of years). For the second tree we created a temporary *phylo* object and rewrote the tip labels with 1 to 4; we will discuss tip labels later. We plot both trees in the same figure using the mfrow argument in par():

```
par(mfrow=c(1,2))
plot(tree.owls,edge.width=3,label.offset=0.3)
add.scale.bar()
text(1,1.1,"MY")

owl.tmp<-tree.owls
owl.tmp$tip.label<-c(1:4)
```

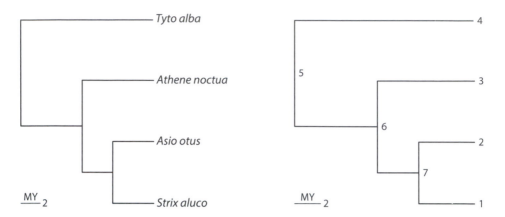

Figure 2.3. Two trees for the same owl phylogeny. The first is the default representation with tips labeled with species names. The second tree shows interior node and tip numbers that correspond to tree.owls$edge (see text).

```
plot(owl.tmp,edge.width=3, show.node.label=TRUE,label.
     offset=0.3)
add.scale.bar()
text(1,1.1,"MY")
```

For large complex trees, it is important to remember the node numbering rules. Number 1 is the bottom tip and number n is the topmost tip, corresponding to n species. Next, $n + 1$ is the root. For all subsequent splits, the next node number is the bottom one. Here is code creating a tree with 6 tips, which is then plotted and the edge matrix returned:

```
tr1<-read.tree(text="(((t2:0.1610994965,t4:0.161
     0994965):0.2636854169,t5:0.4247849134):0.5
     752150866,(t1:0.8616120011,(t6:0.578447259
     8,t3:0.5784472598):0.2831647414):0.1383879989);")
```

```
plot(tr1,show.tip.label=FALSE,edge.width=3)
```

```
tr1$edge
```

In figure 2.4, we can see that the node numbers follow our rule with the tips numbered 1–6, starting from the bottom. Next, the root is numbered 7, and 8 and 9 follow the nodes along the bottom of the phylogeny—when plotted using the default orientation.

Let's go back to our owl tree. There are two other important pieces of information we see when we use `attributes(tree.owls)`, namely edge lengths and tip labels. We can call either of these objects in the same way as in the edge matrix. Let's start with tip labels. We call them using:

```
tree.owls$tip.label
```

This returns:

```
[1] "Strix_aluco"  "Asio_otus"  "Athene_noctua"  "Tyto_alba"
```

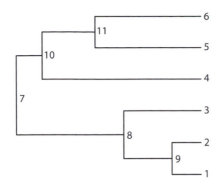

Figure 2.4. A phylogeny showing the default node numbering in *ape*.

These are ordered as 1–4 from the edge matrix, or bottom to top when we plot the tree. In the second tree in figure 2.3, we rewrote the tip labels, which is a very straightforward thing to do; however, when doing so we should be careful that we are ordering the labels correctly. The edge lengths are called in a similar manner:

```
tree.owls$edge.length
```

This returns:

```
[1]  6.3  3.1  4.2  4.2  7.3  13.5
```

These are the edge lengths that correspond to the edge matrix. This vector of edge lengths is the same length as the number of rows in the edge matrix and are ordered the same. Thus, looking back at the owl edge matrix, the first row indicates the edge from node 5 (the root) to 6, and this edge is 6.3 units long (in our case, units are millions of years), and the second row is the edge from node 6 to 7 and it is 3.1 units long, and so on. The longest edge—13.5 units long—corresponds to the last row of the edge matrix, which is from the root (node 5) to tip 4, and so spans the entire length of the tree. Edge lengths can easily be rewritten, just as for the node labels, but again care should be taken when doing this to ensure that lengths and edges line up correctly.

This *phylo* object, which is just a series of matrices and vectors, is logical and straightforward to interpret. This is why so many other packages have *ape* dependencies; that is, they require *ape* to be installed. However, it is important to recognize that the *ape* library itself contains many useful features and functions. Throughout the various chapters in this book, we will utilize many of these functions and explain them in more detail as we go along; here we provide a table that summarizes some of what we think are the most useful and commonly used functions (table 2.1). A more detailed review is provided by the author of the *ape* library, Emmanual Paradis, in his book *Analysis of Phylogenetics and Evolution with R* (Paradis 2011).

2.1.2. Picante

The other package we use extensively in this book is *picante* (Kembel et al. 2010), which provides many of the basic tools to test ecological hypotheses using phylogenies. The functions in *picante* typically use up to three different data types, and they each need to be formatted

TABLE 2.1.

Useful functions contained in *ape*, the main inputs and arguments that need to be specified, and a brief description of their use.

Function	Main inputs and arguments	Description
chronos	*Phylo* object, lambda value	This performs rate smoothing to create an ultrametric tree.
cophenetic.phylo	*Phylo* object	This returns all pairwise distances between species. Just use *cophenetic*(); the *'phylo'* is implied from the object type.
drop.tip	*Phylo* object, vector of tip labels to be dropped	This creates a new tree without tips that are to be excluded.
mrca	*Phylo* object	This returns a matrix of ancestral nodes for all pairs of tips.
multi2di	*Phylo* object	This function resolves polytomies.
pic	*Phylo* object, vector of trait values	This performs phylogenetically independent contrasts, useful for analyses where phylogenetic relationships need to be controlled for.
plot.phylo	*Phylo* object	This plots the phylogeny. Just use *plot*(); the *'phylo'* is implied from the object type.
read.tree	File or text	This reads in a phylogeny and creates a *phylo* object.
rcoal	Number of tips	This returns a random coalescence tree, which is ultrametric.
root	*Phylo* object, outgroup	This roots or reroots a tree.
rtree	Number of tips	This returns a random tree, which is not ultrametric.
vcv.phylo	*Phylo* object	This returns the variance-covariance matrix representing species relatedness. Just use *vcv*(); the *'phylo'* is implied from the object type.
write.tree	*Phylo* object, file	This writes a Newick file of the tree to the specified location.

correctly. The first data type is the phylogeny, and the *picante* functions conveniently use the *ape* "phylo" object, which we have already covered. The second data type is a community data matrix, and *picante* uses the format established by a well-known ecologically orientated multivariate package called *vegan*; this matrix has taxa or species across the columns and sites or samples down the rows (e.g., table 2.2). The cells can be presence-absence or abundances, and the functions in *picante* have options to use the available abundances or just use presences.

The final data type used by *picante* is a trait matrix. This is a taxon by trait matrix with taxa as rows, and traits as columns. Each column is a different trait and so the values in each column can be quite different from one another. Table 2.3 shows an example of what the taxon by trait matrix looks like.

Picante comes with a well-written instruction manual (http://picante.r-forge.r-project .org/picante-intro.pdf) to show how these data types are represented and used. *Picante* also comes with sample data sets that are correctly formatted and will work with the functions.

TABLE 2.2.
A sample community data matrix.

	spA	spB	spC	spD	spE
Site1	0	3	1	4	12
Site2	5	9	2	0	6
Site3	7	5	9	0	9
Site4	0	0	4	2	10
Site5	3	2	10	1	7

TABLE 2.3.
An example taxon by trait matrix.

	Trait1	Trait2	Trait3
spA	26	340	11
spB	18	289	12
spC	37	508	29
spD	30	440	24
spE	32	412	19

Briefly, let's look at one of these. *Picante* includes a basic example, a data set called "phylo-com,"[1] which can be called:

```
library(picante)
data(phylocom)
phylocom
```

This returns a list object with three elements (the names starting with $) that includes the phylogeny, community matrix, and trait matrix we described above. An easier way to see what this object contains is to use the attributes function:

```
attributes(phylocom)
```

This returns:

```
$names
[1] "phylo" "sample" "traits"
```

It shows us that there are three named elements. We can pull out any of these data types by using the $, for example:

```
comm<-phylocom$sample
comm
```

So we now have access to an object called *comm* and we can do the same for the other two:

```
phy<-phylocom$phylo
traits<-phylocom$traits
```

1. The name "phylocom" refers to the *picante* precursor called Phylocom (http://phylodiversity.net/phylocom/).

A number of functions will call for two data types: a phylogeny and either the community or trait matrix. For example, the function calculating phylogenetic distance, PD (see chapter 3), uses the phylogeny directly and includes the following input:

```
pd(comm,phy)
```

and returns:

	PD	SR
clump1	16	8
clump2a	17	8
clump2b	18	8
clump4	22	8
even	30	8
random	27	8

We will discuss this output in chapter 3, but the point is that we were able to calculate something called PD using the community data and phylogeny directly. However, the functions that calculate mean pairwise distance (MPD) and mean nearest taxon distance (MNTD)—covered in chapter 3—require the species pairwise phylogenetic distance matrix, rather than the phylogeny. To calculate MPD we specify the arguments in the function like this:

```
mpd(comm,cophenetic(phy))
```

And this simply returns a vector of mean pairwise distances, one for each sample. The function `cophenetic` was mentioned in table 2.1, and alone it is used like this:

```
cophenetic(phy)
```

It returns a large matrix with the pairwise phylogenetic distances between all pairs of species. We will discuss this in greater detail in chapter 3.

Picante includes a number of useful functions for calculating community phylogenetic patterns (chapter 3), turnover among sites (i.e., beta diversity, chapter 6), and for analyzing trait evolution (chapter 5). There are many more useful functions, which will be explained throughout this book, but those in table 2.4 are most often employed for basic preliminary analyses of community data sets.

2.1.3. Other Packages

There are number of other packages that will be used throughout this book, and we will mention a few here. The two packages above—*ape* and *picante*—are very general packages, with many functions that represent a broad array of phylogenetic manipulations, calculations, and analysis. As mentioned previously, there are a suite of more specialized packages that offer important tools. These include:

1. *pez* (Pearse et al. 2015): This new package provides additional tools and metrics not provided in *picante*. It includes a number of measures mentioned in chapter 3, as well as phylogenetic simulations and methods to combine traits and phylogeny mentioned in chapter 10.
2. *geiger* (Harmon et al. 2008): This package provides many powerful tools to transform and analyze phylogenies, and to perform simulations. We will use *geiger* to

TABLE 2.4.
Useful functions contained in *picante*, the main inputs and arguments that need to be specified, and a brief description of their use.

Function	Main input and arguments	Description
comdist, comdistnt	Community matrix, phylogenetic distance matrix using *cophenetic()*	This returns average pairwise distances between community pairs.
evol.distinct	*Phylo* object	This returns a measure of species evolutionary distinctiveness.
mpd, mntd	Community matrix, phylogenetic distance matrix using *cophenetic()*	These functions return a vector of mean pairwise or nearest taxon distances within communities.
pd	Community matrix, *Phylo* object	This returns community phylogenetic diversity values.
phylosignal	A vector of trait values, *Phylo* object	This returns Blomberg's K and PIC values, both measures of phylogenetic signal. Note that the trait vector needs to be in the same order as the tip labels.
ses.mpd, ses.mntd,	Community matrix, phylogenetic distance matrix using *cophenetic()*	These functions use randomizations to calculate standardized effect sizes and significance of mean pairwise, or nearest taxon distances within communities.
ses.pd	Community matrix, *Phylo* object	This function uses randomizations to calculate standardized effect sizes and significance of community phylogenetic diversity values.

transform trees to reflect different assumptions about evolutionary rates (chapter 4) and to simulate and analyze trait evolution (chapter 5).

3. *vegan* (Oksanen et al. 2008): This package includes numerous functions performing both standard and community multivariate analyses. *Vegan* has become one of the most commonly used packages for community ecologists and includes a number of distance-based calculations, which form the basis of many ecophylogenetic analyses.

2.2. BUILDING TREES

Before we can start analyzing ecophylogenetic patterns, we of course need a phylogenetic tree of the taxa in our sample (and perhaps of the regional species pool). One strategy would be to find a published phylogeny in the primary literature or online databases such as Tree-Base (http://treebase.org/); see section 2.3. However, it is likely that available phylogenetic trees do not contain the exact suite of species contained within study sites, and researchers should therefore consider estimating a phylogenetic tree for their species list. The methods used for phylogenetic reconstruction have engendered many, sometimes heated, debates. Discussion on whether or not to include morphological data or the independence of characters, and the appropriateness of evolutionary models populate the literature; see, for example, J. Felsenstein's *Inferring Phylogenies* (2004) and Z. Yang's *Molecular Evolution: A Statistical Approach* (2014), for an introduction into this literature. Most modern phylogenetic trees are constructed from DNA sequence data, and we will focus on using such data for phylogeny reconstruction here. One of the most impressive biological databases in existence today is that containing the vast and rapidly growing wealth of genetic sequence data. Commonly referred to simply as GenBank, this database is in fact a conglomeration of the DNA DataBank of Japan (DDBJ), the European Molecular Biology Laboratory (EMBL), and GenBank at the National Center for Biotechnology Information (NCBI). Perhaps most remarkably, these data are free to access and use by anyone with an internet connection. Their availability have provided researchers with the potential to reconstruct phylogenetic relationships among a vast group of taxa, and helped facilitate the "phylogenetic revolution" in ecological research.

The generation of stand-alone computational tools for phylogeny reconstruction has become an industry in itself, but R provides us with a simple interface for querying GenBank and for generating increasingly robust estimates of phylogenetic relationships. However, the reader will likely want to explore other programs, depending on the size of their data set (number of taxa) and research goals. (Questions asked might include whether a dated phylogeny is necessary, whether nodal support values are important, whether a single consensus tree is desirable, whether it is important to explore a set of equally possible tree topologies, etc.). The software that dominated the field through the 1990s and into the 2000s was PAUP* (Phylogenetic Analysis Using Parsimony[*and other methods]), written by D. Swofford (2003). As suggested by the name, PAUP was largely written to optimize tree searches using parsimony, although later versions also included distance and likelihood methods. Likelihood approaches, which include explicit evolutionary models (it should be noted that parsimony also represents a particular evolutionary model), were attractive because they could, at least theoretically, account for evolutionary reversals and multiple substitution events, but they were computationally intensive and difficult to use. However, the field developed at a pace spurred by developments that moved the computational load from researcher desktops to remote servers. Although it is not possible to provide a full review of all the tools available here, we have picked out PhyML (http://atgc.lirmm.fr/phyml/; Guindon et al. 2010), and RaxML (http://sco.h-its.org/exelixis/web/software/raxml/index.html; Stamatakis 2014) for particular mention because they are both commonly used and well supported. With increased processing speed, new Bayesian methods that implement Markov chain Monte Carlo (MCMC) methods for sampling from a probability distribution are now becoming more common, allowing for the sampling of parameter space representing millions of trees. Two of the most widely used implementations are MrBayes (http://mrbayes.sourceforge.net/; Ronquist and Huelsenbeck 2003) and BEAST (http://beast.bio.ed.ac.uk/; Drummond et al. 2012). The particular demands of these algorithms are not always best suited to the R environment (although you may be able to call these packages from R using some basic programming skills), so we do not expand on these methods here, but refer the reader back to the primary literature (Huelsenbeck and Ronquist 2001, Archibald et al. 2003, Holder and Lewis 2003). Here we provide the basic steps for reconstructing a phylogeny in R from sequence data, with a focus on ML methods. We assume the reader has a list of taxa for which they wish to reconstruct the phylogeny, and walk through how to query GenBank and download raw sequence data, sequence alignment, and phylogenetic inference (e.g., fig. 2.5).

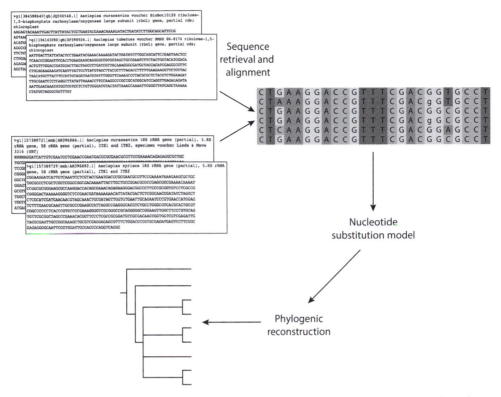

Figure 2.5. Schematic representing the various steps required to generate a phylogenetic hypothesis from DNA sequence data as described in the following sections.

2.2.1. Downloading and Aligning DNA Sequences

The first step in reconstructing phylogenetic relationships is to generate a species matrix for the gene sequences of interest. Given a list of GenBank accession numbers we can easily use the `read.GenBank()` function in *ape* to extract sequence data, here using the *rbcL* sequence for the wild oat, *Avena fatua*, as an example:

```
Avena.fatua<-read.GenBank("AJ746257.1")
Avena.fatua
1 DNA sequence in binary format stored in a list.

Sequence length: 1428

Label: AJ746257.1

Base composition:
      a       c       g       t
  0.279   0.189   0.247   0.285
```

This function returns a *DNAbin* object containing the sequence data in binary format. We are told the sequence length is 1,428, which is reassuring as this is the expected length for complete *rbcL* sequences. We should remember, however, that many sequences in GenBank might be of different lengths; shorter reads are frequent due to the recent popularity of DNA barcoding, which amplifies shorter gene fragments, and in some cases researchers may have sequenced beyond the gene. Should we wish to examine the sequences of bases directly, we can view them by converting the *DNAbin* object into a character string:

```
as.character(Avena.fatua)
```

More commonly, we might have a list of species names but no information on the accession numbers matching them to the relevant DNA sequences in GenBank. Fortunately, the query() function in the *seqinr* library provides a more powerful tool for interfacing with GenBank, and allows us to search for particular species, organelles, and named genes (as long as the GenBank annotations are accurate). Using this function, we first open a connection with GenBank, choosebank("genbank"), and then submit our query; last, to keep things tidy, we then close the connection, closebank():

```
choosebank("genbank")
Avena.fatua <-query("Avena.fatua", "sp=Avena fatua AND
     K=rbcL")
getSequence(Avena.fatua$req[[1]], as.string = TRUE)
closebank()
```

Here we have submitted a query specifying both the species Latin binomial (*Avena fatua*) and the gene (*rbcL*). We then extract the sequence directly using getSequence(); if we specify as.string = TRUE we return a character string composed of the base sequences. In some cases GenBank might contain multiple copies of a sequence for the same species; in the example above we have arbitrarily selected the first entry in the list using the double bracket notation [[1]] in the getSequence() function. However, we might want to alter this criterion—for example, by choosing the longest read, the sequence that is closest to the known length of the gene, or other options.

Because we are interested in the relationships among many species, we need to repeat this procedure for each species and can thus write a short loop to generate a molecular matrix for a given species list. For illustration, we use the species list from the annual grassland plant community at Stanford University's Jasper Ridge biological preserve, which we return to in subsequent chapters. For simplicity, we have only included species for which we have complete Latin binomials, and for which we know *a priori* that there is matching sequence data in GenBank. In the example below, we first read in the species list for Jasper Ridge and create an empty list object *dna.matrix* in which to store our data. We then run our GenBank query on each species in our list, using rbind() to generate our matrix, and adding a row to our matrix for each new species:

```
species<-read.csv("Jasper species.txt", header=F)
dna.matrix<-list()

choosebank("genbank")

for (x in 1:length(species[,1])){
my.query<-paste("sp=", species[x,1]," AND K=rbcL", sep="")
jasp <-query("jasp", my.query)
dna.matrix[[x]]<-getSequence(jasp$req[[1]])
}

closebank()
```

We are well on our way to collating the data we need to reconstruct a phylogeny; however, if you inspect the matrix you will notice that the sequences are of different lengths and mis-aligned. We therefore still need to go through a few more steps before we are ready to start generating trees. Before continuing, we note that researchers might find it more efficient to query GenBank directly—for example, through the NCBI GUI. GenBank allows users to download sequence data in various formats; perhaps most common is the FASTA format, which was derived from the FASTA software package by D. Lipman and W. Pearson (Lipman and Pearson 1985, Pearson and Lipman 1988). We can use *seqinr* to read in FASTA files directly, which in this example is the FASTA file for *Avena fatua* that we queried GenBank for above:

```
fasta.avena<-read.fasta(file = "Avena.fatua.fasta", as.string
    = TRUE, seqtype = "DNA")
```

For our purposes, however, we are interested in writing our *DNAbin* object containing all our sequences for the Jasper Ridge community that we created above, to a FASTA file that we can import into our favorite alignment program:

```
write.fasta(dna.matrix, as.vector(species[,1]), file.out=
    "Jasper.fasta")
```

Aligning sequences is a deceptively difficult task; fortunately, there are many programs available to choose from. Some of the most common freeware include Clustal (http://www.clustal.org/; Larkin et al. 2007), MAFFT (http://mafft.cbrc.jp/alignment/software/; Katoh et al. 2002) and MUSCLE (http://www.drive5.com/muscle/; Edgar 2004). Of course it is possible to call any command-line program from R (assuming it is already installed). For example, if we wish to align our sequences using ClustalW2 (the command-line version of Clustal), we can simply call the program as follows (remember to specify to the correct working directory in R and to place the FASTA file in the Clustal root directory):

```
system(paste('"./clustalW2" Jasper.fasta')
```

We can read this file back into R using `read.dna()`:

```
Jasper.aln<-read.dna("Jasper.aln", format = "clustal")
```

We can alternatively implement the MUSCLE algorithm directly within R using the package *muscle*, and save it as a FASTA file.

```
Jasper.muscle <-muscle(seqs = "Jasper.fasta")
write.fasta(Jasper.muscle, file = "Jasper.muscle.fasta")
```

Most algorithms perform reasonably well on length-conserved coding regions, such as *rbcL*, which we use here. Noncoding regions and sequences with multiple insertions and deletions (indels) can be more problematic. Irrespective of marker, it is good practice to visually check alignments. We can use the `print()` function to display our *muscle* alignment, here displaying the first 100 bases (fig. 2.6):

```
print(Jasper.muscle, from = 1, to = 100)
```

We can also visualize our alignment in *ape* (fig. 2.7) using `image.DNAbin()`, first converting our *muscle* file to a `DNAbin` object, for which we need to install the *adegenet* library:

```
Jasper.DNAbin<-fasta2DNAbin("Jasper.muscle.fasta",
    chunkSize=10)
```

```
# 24 sequences with 1446 positions

# Position 1 to 100:

                           1                                                                                                    100
Brachypodium_distachyon    ATGTCACCACAAACAGAAACTAAAGCAAGTGTTGGATTTAAAGCTGGTGTTAAAGATTATAGATTGACTTACTACACCCCGGAGTATGAAACCAAGGATA
Elymus_glaucus             -TGTCACCACCAACAGAAACTAAAGCAGGTGTTGGATTTCAAGCTGGTGTTAAAGATTATAAATTGACTTACTACACCCCAGAGTATGAAACTAAGGATA
Bromus_diandrus            ------------------------CTGGTGTTAAAGATTATAAATTGACTTACTACACCCCAGAGTATGAAACTAAGGATA
Bromus_hordeaceus          --------------------TAAAGCAGGTGTTGGATTTCAAGCTGGTGTTAAAGATTATAAATTGACTTACTACACCCCAGAGTATGAAACTAAGGATA
Aira_caryophyllea          ------------------------AAGTGTTGGATTTCAAGCTGGTGTTAAAGATTATAAATTGACTTACTACACCCCGGAGTATGAAACCAAGGATA
Lolium_multiflorum         --------------------CTAAAGCAAGTGTTGGATTTCAAGCTGGTGTTAAAGATTATAAATTGACTTACTACACCCCGGAGTATGAAACCAAGGATA
Vulpia_bromoides           ----------------------------CTGGTGTTAAAGATTATAAATTGACTTACTACACCCCGGAGTATGAAACCAAGGATA
Vulpia_myuros              ------------------------CTAAAGCAAGTGTTGGATTTCAAGCTGGTGTTAAAGATTATAAATTGACTTACTACACCCCGGAGTATGAAACCAAGGATA
Briza_minor                -----------------------------CTGGTGTTAAAGATTATAAATTGACTTACTACACCCCGGAGTATGAAACCAAGGATA
Avena_barbata              ----------------------------CTGGTGTTAAAGATTATAAATTGACTTACTACACCCCGGAGTATGAAACCAAGGATA
Avena_fatua                ATGTCACCACAAACAGAAACTAAAGCAAGTGTTGGATTTCAAGCTGGTGTTAAAGATTATAAATTGACTTACTACACCCCGGAGTATGAAACCAAGGATA
Geranium_dissectum         -----------------------CGGGTGTTAAAGACTATAAATTGACTTATTATACTCCTGATTATGAAACCAAGGATA
Melilotus_indica           ------------------------CTGGTGTTAAAGATTATAAATTGACTTATTATACTCCTGAGTATGAAACCAAAGATA
Vicia_sativa               ------------------------------AGATTATAAATTGACTTATTATACTCCTGAATATGAAACCAAGGATA
Anagallis_arvensus         ATGTCACCACAAACAGAGACTAAAGCAGGTACTGGATTCAAAGCTGGTGTTAAAGATTACAAATTGACTTATTATACTCCTGAGTATGTAACCAAGGATA
Torilis_nodosa             ------------------------CTGGGTTAAAGATTACAAATTGACTTATTATACTCCTGAGTATGAAACCAAGGATA
Crepis_vesicaria           --------------------AAAGCAAGTGTTGGATTCAAAGCTGGTGTTAAAGATTATAAATTGAATTATTATACTCCTGAGTATGAAACCAAGGATA
Filago_gallica             ------------------------CTGGTGTTAAAGATTATAAATTGACTTATTATACTCCTGAGTATGAAACCAAGGATA
Hypochaeris_radicata       --------------------TAAAGCAAGTGTTGGATTCAAAGCTGGTGTTAAAGATTATAAATTGACTTATTATACTCCTGAATATGAAACCAAGGATA
Lactuca_serriola           ------------------------AAGTGTTGGATTCAAAGCTGGTGTTAAAGATTATAAATTGACTTATTATACTCCTGAGTATGAAACCAAGGATA
Silybum_marianum           ---------------------------------------------------------ATTATACTCCTGAGTATGAAACCAAGGATA
Sonchus_asper              ------------------------AAGTGTTGGATTCAAAGCTGGTGTTAAAGATTATAAATTGACTTATTATACTCCTGAGTATGAAACCAAGGATA
Convolvulus_arvensis       --------------------ACTAAAGCAAGTGTTGGATTCAAAGCTGGTGTAAAAGCTTACAAATTGACTTATTATACTCCTGAGTACGAAACCAAAGATA
Sherardia_arvensis         ----------------------------------------------------------ATTATACTCCTGAATACGAAACCAAAGATA
```

Figure 2.6. Example of aligned *rbcL* sequences for Jasper Ridge, showing the first 100 bases.

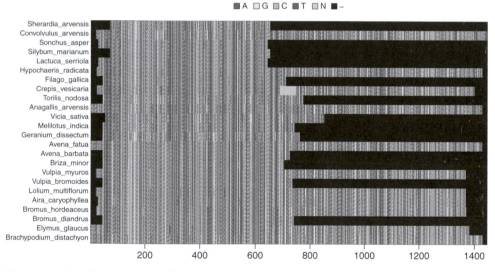

Figure 2.7. Visualization of complete sequence alignment shown in figure 2.6, with bases identified by shading.

In this example we have reduced the font size of the taxon labels to the left of the alignment using `cex.lab` to improve readability. The argument `chunkSize` specifies the number of sequences read at a time; increasing this value decreases the computational time required to read the data file but increases memory requirements. We can see clearly that the sequences are of different length, and a number of base substitutions are evident. Depending on your preferences, you might prefer to visualize the alignment in another program; this could be useful if the sequences need to be manually edited. Once we are happy with our alignment we are ready to reconstruct our phylogeny.

2.2.2. Phylogeny Reconstruction

As we discuss above, there are a plethora of tree-building programs and algorithms. It is impossible here to provide a comprehensive overview; rather, we illustrate some different tree-building approaches that are easily implemented in R, from the simplest distance-based methods to more complex maximum likelihood (ML) approaches.

2.2.2.1. Distance-Based Methods

Distance-based methods generate a phylogenetic tree from a distance matrix calculated from the aligned sequences, usually clustering taxa using a neighbor-joining (NJ) method. The NJ algorithm starts with a star phylogeny, and sequentially clusters tips based upon their sequence similarity (fig. 2.8). The method is fast, a highly desirable property when analyzing large data sets, and simple to implement. Here we use the `njs()` function in *ape*, which allows for missing sequence data (compare `nj()`), on the DNA distance matrix estimated from our *DNAbin* object generated above:

```
nj.tree<-njs(dist.dna(Jasper.DNAbin, model = "K80"))
plot(nj.tree)
```

The distance matrix is generated using the function `dist.DNA()`, which allows us to specify different models of evolution. In this example we have set `model = K80`, for which transitions and transversions are able to assume separate probabilities. Choice of model can quite dramatically influence the distance calculations, and thus the reconstructed phylogeny. Faced with an array of possible models, evolutionary biologists sought for a way

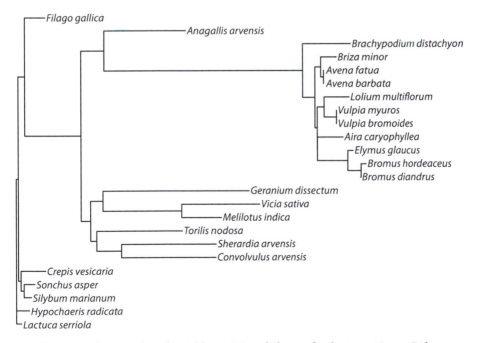

Figure 2.8. Distance-based, neighbor-joining phylogeny for the taxa at Jasper Ridge.

to select among them, and in 1998 D. Posada and K. Crandall published MODELTEST (Posada and Crandall 1998), a simple program that allowed users to compare the fit of alternative models of DNA evolution. As an indication of the breakthrough this provided in the field of phylogenetics, in 2014 the journal *Nature* ranked this paper number 76 among the all-time top-cited papers (Van Noorden et al. 2014), with a whopping 14,099 citations and counting. A version of MODELTEST is implemented in the *phangorn* library, and requires a *phyDat* object containing our aligned sequence which we can generate from our FASTA file above:

```
Jasper<-read.phyDat("Jasper.muscle.fasta", format = "fasta",
    type = "DNA")
modelTest(Jasper, tree=NULL, model = c("JC", "K80"),
    G = FALSE, I = FALSE, k = 1)

    Model  df      logLik        AIC        BIC
    1JC     45   -5626.24   11342.48   11579.92
    2K80    46   -5501.61   11095.22   11337.94
```

Here we have compared the Jukes Cantor model, in which all substitutions have the same probability, to our K80 model. We can see clearly that the K80 is favored by both AIC (Akaike Information Criterion) and BIC (Bayesian Information Criterion), and thus our earlier choice of this more complex model (and one additional parameter) was justified. However, there is no need to stop here, and we can evaluate fits for increasingly parameter-rich models, including a gamma parameter to describe rate variation across sites (G = TRUE), a proportion of invariant sites (I = TRUE), and the number of rate classes (K = N), the default being K = 4. It is also worth noting that MODELTEST requires a tree to evaluate model fits; fortunately for us this tree does not have to be very accurate, and we left this option as the default (tree=NULL). However, the default was to assume a NJ tree (at least in previous iterations of MODELTEST), resulting in some circularity when using an NJ method to generate your phylogeny. Usually we wish to explore more sophisticated tree-building algorithms and we can thus take our outputs from MODELTEST to help parameterize them.

While distance-based methods have advantages, they have been criticized because they rely on the accuracy of the underlying distance matrix, which cannot be known with confidence, and return only a single tree. In addition, under some circumstances NJ methods can result in negative branch lengths, a particularly undesirable feature for phylogenetic trees.

2.2.2.2. Parsimony

Parsimony methods are conceptually simple, but often computationally challenging due to the vastness of tree space. The number of possible tree topologies increases exceptionally fast with the number of taxa; for 3 taxa there are just 3 possible fully resolved rooted trees, for 4 taxa 15 trees, for 5 taxa 105 trees, and for just 10 taxa there are over 34 million possible rooted trees (Felsenstein 1978). The principle of phylogeny reconstruction using maximum parsimony is to find the tree topology that minimizes the number of evolutionary steps required to explain the data. Because it is impossible to compare all possible trees, various algorithms have been developed to guide this search, and these differ in details on how tree space is traversed. Parsimony searches can be implemented using the *phangorn* library. We again use our Jasper Ridge example, and the *phyDat* object we generated above. To help the analyses along we provide a starting tree, here our NJ tree, and the parsimony algorithm then proceeds by rearranging branches on the tree, using either Nearest Neighbor Interchange (NNI) or sub tree pruning and regrafting (SPR). The latter tends to search tree space more thoroughly, reducing the probability that the search will get stuck on a local optima, but this can also increase run times:

```
pars.tree<-optim.parsimony(nj.tree, Jasper,
     rearrangements="SPR")
```

We can then evaluate the length of the tree (number of inferred evolutionary substitutions):

```
parsimony(pars.tree, Jasper)
```

Remember, under parsimony the shorter the length of the tree the better. The parsimony ratchet of Nixon (1999), is typically more efficient at searching tree space, and can often return shorter (more parsimonious) trees, especially in large data sets:

```
pratch.tree<-pratchet(Jasper, start=nj.tree,
     rearrangements="SPR")
parsimony(pratch.tree, Jasper)
```

Now let us plot both trees (fig. 2.9), with branch lengths estimated using the ACCTRAN (accelerated transformation) optimization, which favors reversals over parallelisms when the choice is equally parsimonious, and will thus tend to push evolutionary events toward the root (Wiley and Collins 1991):

```
pars.tree<-acctran(pars.tree, Jasper)
pratch.tree<-acctran(pratch.tree, Jasper)

par(mfrow=c(1,2))
plot(midpoint(pars.tree))
title("parsimony")
plot(midpoint(pratch.tree))
title("parsimony ratchet")
```

Remember that we have not yet specified an outgroup to root the tree, so we have here arbitrarily specified midpoint rooting to better compare topologies, in which the root is placed at the midpoint along the longest path separating any two taxa on the tree. We can see that the two topologies differ slightly, and in this example the parsimony ratchet finds a slightly shorter tree. It is tempting, therefore, to assume that the ratchet tree is the more accurate; however, there are often a very large number of equally parsimonious trees (trees of the same length), and it can be misleading to select any one tree. Further, assumptions of parsimony are often violated in the real world where homoplasy (i.e., convergent evolution, parallel evolution, and evolutionary reversals) is rife, and it has thus been argued that focus on the most parsimonious tree(s) is misplaced. Likelihood methods allow us to better account for the contingencies of evolution by the incorporation of a specific evolutionary model into our search algorithms; these are now generally favored over parsimony.

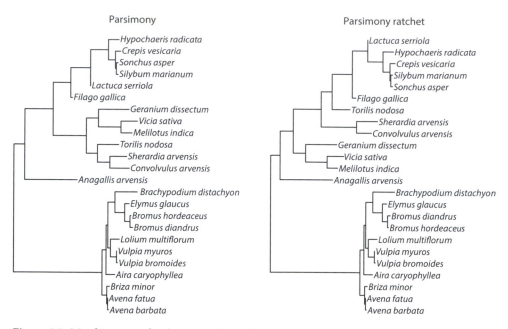

Figure 2.9. Matching trees for the taxa at Jasper Ridge using standard parsimony methods and the parsimony ratchet.

2.2.2.3. Maximum Likelihood (ML)

Likelihood methods evaluate the likelihood of observing the data (aligned sequences) given a tree and a model of evolution. You might note that we are assessing the likelihood of the data rather than the tree, which seems backward, and this is one additional reason that Bayesian approaches are sometimes preferred. Nonetheless, as in parsimony approaches, likelihood algorithms proceed by searching tree space; but rather than comparing tree lengths they compare likelihoods. First we again generate a starting tree using a neighbor-joining method; this time we generate our NJ tree using *phangorn* (note the upper case NJ() function) on a distance matrix calculated using the function dist.logDet() (similar to the option model = "BH87" in the dist.DNA() function within *ape*) on the *phyDat* object. This allows for asymmetric rates of nucleotide change.

```
dist.jasper<-dist.logDet(Jasper)
nj.tree<-NJ(dist.jasper)
```

The likelihood of the tree given the data can be calculated as follows:

```
ml.model<-pml(nj.tree, Jasper)
ml.model

loglikelihood: -14771.56

unconstrained loglikelihood: -6165.087

Rate matrix:
  a c g t
a 0 1 1 1
```

```
c 1 0 1 1
g 1 1 0 1
t 1 1 1 0
Base frequencies:
0.25 0.25 0.25 0.25
```

Now we can optimize our tree using function `optim.pml()`, here sticking with our K80 substitution model, and then compare log likelihoods:

```
ml.tree<-optim.pml(ml.model, model = "K80")
ml.tree

loglikelihood: -6147.331

unconstrained loglikelihood: -6165.087

Rate matrix:
  a c g t
a 0 1 1 1
c 1 0 1 1
g 1 1 0 1
t 1 1 1 0

Base frequencies:
0.25 0.25 0.25 0.25
```

We can see a significant improvement in model fit (less negative log likelihoods); however, by default only branch lengths are optimized. Now let us simultaneously optimize the tree topology by setting `optNni = TRUE`:

```
optim.ML.tree<-optim.pml(ml.model, model = "K80",
        optNni=TRUE)
optim.ML.tree

loglikelihood: -5437.872
unconstrained loglikelihood: -6165.087

Rate matrix:
    a c g t
  a  0.000000  1.000000  3.768566  1.000000
  c  1.000000  0.000000  1.000000  3.768566
  g  3.768566  1.000000  0.000000  1.000000
  t  1.000000  3.768566  1.000000  0.000000

Base frequencies:
0.25  0.25  0.25  0.25
```

Unsurprisingly, we get a further improved log likelihood. We can plot the ML tree as before (fig. 2.10) and compare it to our estimates using NJ and parsimony:

```
plot(midpoint(optim.ML.tree$tree))
title("optimised ML tree")
```

We discussed above model selection using `modeltest()`, and `optim.mpl()` can implement any of the evolutionary models selected. Additional parameters allow for

Optimized ML tree

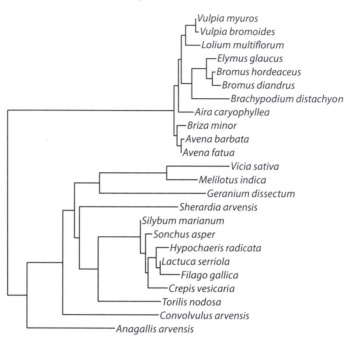

Figure 2.10. Maximum Likelihood tree for the Jasper Ridge data assuming the K80 model of nucleotide substitution.

fine-tuning the search algorithm, and are detailed in the help pages. We have here described a simple case scenario, using a single coding gene. However, it is possible to partition the alignment into separate gene regions, and fit separate evolutionary models to each partition (see `pmlPart()`).

2.2.2.4. Bayesian Inference

Although we do not explore the implementation of Bayesian inference in phylogeny reconstruction within R, its recent popularity in the tree reconstruction literature should be recognized. Bayesian inference is derived from Bayes' theorem, an expression of inverse probability developed by the Reverend Thomas Bayes and published after his death in 1763. The philosophy of Bayesian approaches is appealing and is closely allied to ML, but here we evaluate the likelihood of the tree given the data, which seems sensible since the data is our observation. In addition, Bayesian methods allow complex, parameter-rich, models of sequence evolution to be implemented. Complex models can be difficult to fit using traditional ML approaches, especially when the ratio of data points to parameters is low, and ML estimates of parameters can become unreliable.

Bayesian methods do have some drawbacks, however. First, they are computationally intensive, and the application of Bayesian inference in phylogenetic reconstruction did not take off until the introduction of Monte Carlo Markov Chain (MCMC) methods in the 1990s, which allowed for the rapid search of parameter space. Nonetheless, it can still be difficult to determine whether the MCMC algorithm has achieved stationarity, and often many millions of "generations" are required to be run. Second, Bayesian inference requires specification of priors, which are often unknown, but which might still influence the posterior distribution. Priors indicate our beliefs in particular outcomes before having seen the data. It is often argued that uninformative priors (which assign equal probabilities to all possibilities) can be specified, such that the posterior distribution will be largely guided by the data; although by doing this we of course violate Bayesian philosophy.

We refer the reader to a short review paper by Holder and Lewis (2003) that provides a succinct comparison of tree reconstruction methods.

2.3. FINDING AND ADAPTING AVAILABLE TREES

In many cases, especially for well-studied taxa or locations, a phylogeny might already exist for the group of interest and we will not want to waste time duplicating efforts. If systematists familiar with the group generated the published phylogenetic tree, we might also have increased confidence in the phylogenetic placement of included taxa. Unfortunately, the sharing of phylogenetic trees lags behind the sharing of genomic data, and there is no single repository that we can search for all published phylogenies, despite growing calls for better sharing and reuse of published phylogenies (Stoltzfus et al. 2012). The technical challenges in the generation of such a database are not insurmountable; however, some efforts have already fallen by the wayside. For example, funding for the PANDIT (Protein and Associated Nucleotide Domains with Inferred Trees) database, originally hosted by the European Bioinformatics Institute (http://www.ebi.ac.uk/) was frozen in 2008. It might, therefore, be unsurprising to learn that the function for querying this database through R (previously implemented in the *apTreeshape* library) is now depreciated. Fortunately, TreeBASE (http://treebase.org), another repository of phylogenetic data that launched in 1994 (Sanderson et al. 1994) is still active and growing, with over 12,000 phylogenies at the last count. While the proportion of studies submitting data to TreeBASE remains relatively small, it is easily queried through R using the library *treebase*[2] (Boettiger and Temple Lang 2012). In this example we search for phylogenies including the taxon *Watsonia*, a genus in the Iris family (Iridaceae) from southern Africa:

```
Iridaceae <-search_treebase('"Watsonia"', by="taxon",
      max_trees=4)
summary(Iridaceae)
```

Currently, TreeBASE contains two studies (although we might hope that this grows over time), including *Watsonia*. The returned object is a list of trees, which can be indexed using the [[ ]] notation (fig. 2.11).

```
plot(Iridaceae[[1]])
```

Alternatively, if we know the author(s) of the study, we can query TreeBASE by author name, here retaining just the first entry,

```
Goldblatt<-search_treebase("Goldblatt", by = "author",
      exact_match = T)[[1]]
```

2. Addendum: at the time of publication the treebase library was no longer regularly supported, and functionality may be reduced.

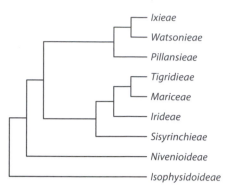

Figure 2.11. Phylogenetic tree for iridaceae retrieved directly from TreeBASE.

Otherwise we can query directly on the unique study identifier in TreeBASE:

```
study.ID <-search_treebase("325", by="id.study")
```

One other potential source of phylogenetic data is the DRYAD digital data repository (http://datadryad.org), which assigns Digital Object Identifiers (DOIs) to data, allowing for better curation and citation by users. The *treebase* library also allows us to query the DRYAD metadata archive using the function `dryad_metadata()` if the study identifier is known. There are of course other resources that provide semiautomated approaches for tree reconstruction outside of the R environment. In the field of plant community ecology, the simplest and perhaps most widely used is the online tool Phylomatic (http://phylodiversity.net/phylomatic/; Webb and Donoghue 2005). In this implementation a fixed backbone tree representing the taxonomic relationships among major lineages is assumed, and missing taxa are then stitched on to this backbone at the minimally inclusive node dictated by taxonomic relationship. Advantages of this approach are that it is very quick, and thus simple, to generate very large phylogenetic trees including many thousands of taxa. However, without additional steps (e.g., manually resolving relationships based on independent phylogenetic hypotheses), output trees are often poorly resolved and do not include branch length information. Depending on your goals, a "phylomatic" tree might be perfectly adequate, but the user should be aware of potential biases when deriving tree metrics from such topologies.

The Open Tree of Life project (https://tree.opentreeoflife.org) provides another potential source of phylogenetic information and is an attempt to work toward a single tree of life for all species by synthesizing published phylogenies using supertree methodologies. The Open Tree of Life is therefore an ambitious attempt to build a single supertree for all life. Supertrees do have limitations; for example, they can be poorly resolved when sources trees are conflicting, and often lack branch length information (see review by Bininda-Emonds et al. 2002). However, the Open Tree of Life website lists contributing source trees, allowing us to refer back to the original publications containing phylogenetic data for particular taxonomic groups.

An exciting new development is the release of new tools that automate the entire process from sequence retrieval to tree reconstruction (fig. 2.5). Phylogenerator (Pearse and Purvis 2013) provides an example of one such approach and offers much promise for the future. A recent review paper by Roquet et al. (2013) provides a useful overview of the various challenges and hurdles faced by such approaches, and also details the path forward.

2.4. TREE SCALING AND RATE SMOOTHING

As we saw in section 2.2, there are a number of approaches for phylogeny reconstruction, and the information they contain may have different meanings. Depending on the information being used to create the tree, edge lengths may represent time, nucleotide substitutions, character changes, or they may be uninformative. Time-trees are generally ultrametric—an ultrametric tree has equal root to tip lengths for all lineages and has a symmetrical pairwise distance matrix (see fig. 2.12)—whereas trees with branch lengths measured in nucleotide substitutions are generally nonultrametric.

The code to create the three trees in figure 2.12 is shown here. We use a random start tree produced by `rtree()` with 12 tips, and so repeating this code will create different trees:

```
tr1<-rtree(12)
tr2<-chronos(tr1)
tr3<-tr2
tr3$edge.length<-NULL

par(mfrow=c(1,3))
plot(tr1,edge.width=2,cex=2, main="A) Non-ultrametric",cex.
     main=2)
plot(tr2,edge.width=2,cex=2, main="B) Ultrametric",cex.
     main=2)
plot(tr3,edge.width=2,cex=2, main="C) Uninformative",cex.
     main=2)
```

All three trees in figure 2.12 originate from the same random tree but they look quite different from one another. In the "uninformative" tree (C), there are no branch lengths (we have specified `tr3$edge.length<-NULL`), and `plot.phylo()` simply draws the tree such that the sister taxa (e.g., *t11* and *t2, t1* and *t6, t10* and *t8*, and *t7* and *t4*) all appear to have the same distance separating them from their common ancestors. Ecologically, the different trees represent evolutionary distances among species quite differently. For example, look at species *t11* and *t2* in the three panels of figure 2.12. The ultrametric version stretches

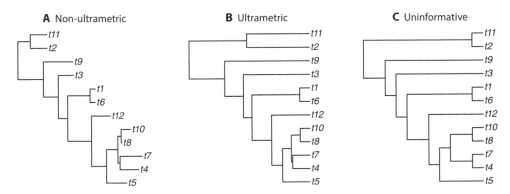

Figure 2.12. Examples of three trees with (*A*) nonultrametric, (*B*) ultrametric, and (*C*) uninformative edge lengths. In the last example, *ape* displays the tree such that it appears ultrametric, but the `$edge` matrix contains no information.

their distance from a common ancestor, and they appear quite distantly related in (B) but closely related in (A). If we were using these trees as evolutionary assumptions to test ecological hypotheses, then tree (B) assumes that there is the potential for a large ecological difference between species *t11* and *t2*, whereas (A) assumes that these species should be relatively ecologically similar. Ecologists need to carefully consider the tree they are using for hypothesis testing and carefully link the tree they have to specific evolutionary assumptions (Davies and Buckley 2012) or transform the edge lengths to explicitly test different assumptions about evolution. Here we will cover some basic methods to transform edge lengths.

A number of the metrics that we cover in chapter 3 assume that the phylogeny is ultrametric (and often that edge lengths are proportional to time), but most phylogenies created from molecular data are not ultrametric. Thus we need to employ meaningful methods to make them ultrametric.

There are much more sophisticated ways to date a phylogeny, but these are beyond the scope of this book. If we wish to use fossils to date individual nodes, we could use a Markov chain Monte Carlo Bayesian analysis to infer likely tree edge distances. If you wish to do this, look at programs such as BEAST (http://beast.bio.ed.ac.uk/), r8s (http://hydrodictyon.eeb.uconn.edu/eebedia/index.php/Phylogenetics:_r8s_Lab), or PhyloBayes (http://www.atgc-montpellier.fr/phylobayes/); we mention some of these approaches in the section on phylogeny reconstruction, above. We cover here some simple functions implemented in R.

2.4.1. Ultrametric Rate Smoothing

A nonultrametric tree implicitly captures different rates of evolution. In figure 2.12A, the clade that includes the taxa *t10*, *t8*, *t7*, *t4*, and *t5* appears to have experienced more evolutionary change than the small clade with *t11* and *t2*. In transforming this tree to an ultrametric one, we wish to "smooth out" these rate differences. Here we will examine two methods. The first uses a mean path length (MPL) method that assumes random mutation and a fixed molecular clock (Britton et al. 2002). The MPL method calculates the age of a node as the mean of all the distances from this node to all tips descending from it. To run this, we use the *ape* function `chronoMPL()` and first we create a random tree with seven tips and then transform the edge lengths:

```
tr.ran<-rtree(7)
tr.ran.u<-chronoMPL(tr.ran)
```

Let's plot these two (and leave a spot open for a final tree):

```
par(mfrow=c(1,3))
plot(tr.ran,edge.width=2,cex=2, main="A) Non-ultramet-
    ric",cex.main=2)
plot(tr.ran.u,edge.width=2,cex=2, main="B)
ChronoMPL",cex.main=2)
```

This produces the first two panels in figure 2.13. The MPL method is fast and easy to implement, but has an unfortunate habit of returning some negative branch lengths. Thus, it should be used with caution.

The other function we use is `chronos`, which uses the penalized maximum likelihood method to estimate divergence times developed by Sanderson (Sanderson 2002). Here is a simple, naive use:

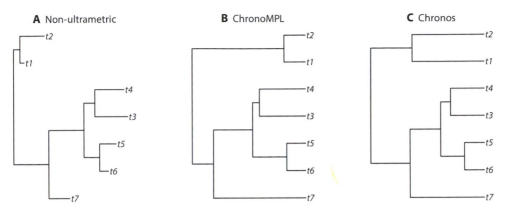

Figure 2.13. There are number of ways to use a rate-smoothing algorithm to transform a nonultrametric tree (A) into an ultrametric one. The first ultrametric tree (B) is transformed using a constant mean path length method, while the second (C) uses a penalized maximum likelihood method.

```
tr.ran.rs<-chronos(tr.ran)
plot(tr.ran.rs,edge.width=2,cex=2, main="C)
Chronos",cex.main=2)
```

This produces figure 2.13C, and it is clear that the two methods produce slightly different (i.e., compare the distances separating *t1* and *t2*), but still ultrametric, trees.

Above, we used the default settings in `chronos()`, but we can optimize the rate-smoothing parameter, λ. A λ of 0 means that each edge can have a unique evolutionary rate, while larger λ values constrain edges to the same rates (Sanderson 2002). We can look at the attributes of the chronos tree in figure 2.13:

```
attributes(tr.ran.rs)
$names
[1] "edge" "tip.label" "edge.length" "Nnode"

$class
[1] "chronos" "phylo"

$order
[1] "cladewise"

$call
chronos(phy = tr.ran)

$ploglik
[1] -9.881871

$rates
[1] 1.6041654 1.4791078 1.9008253 1.9221109 1.9005756
       1.9529343 2.0477430 2.1154029 2.0932130 0.9910776
       0.7541855
[12] 1.0129065
```

```
$message
[1] "relative convergence (4)"

$PHIIC
$PHIIC$logLik
[1] -9.503595

$PHIIC$k
[1] 17

$PHIIC$lambda
[1] 1

$PHIIC$PHIIC
[1] 53.66778
```

We then see that it returns a lot of information not included in the original tree. These additional elements in the list object are returned from the implementation of `chronos`. Let's focus on three of them. The first is `$ploglik`, which is the penalized log likelihood value we will use to select the optimal λ value; second is `$rates`, which gives the rate transformation for each edge in the original phylogeny and is the same length and order as the vector of edge lengths; and finally the third, `$PHIIC$lambda`, gives the λ value the rates were calculated with. Now let's find the optimal λ value using a "for" loop,[3] but first we need to specify a gradient of λ values to test:

```
l<-c(0:10000)
```

Running the function across this many λ values will take some time. Next, let's specify an empty object to act as a container to hold the log likelihood values:

```
LL.out<-NULL
```

Right now this container is empty and without a class. This will be specified once our loop adds a number to it; then R will assume it is a numeric vector. Now let us run our function within the "for" loop:

```
for (i in 1:length(l)){
LL.out[i]<-attributes(chronos(tr.ran,lambda=l[i]))$ploglik
}
```

Now our container is holding 10,001 penalized log likelihood values and we can select the maximum penalized log likelihood value and then the corresponding lambda:

```
l[LL.out==max(LL.out)]
```

This returns a value of 0, corresponding to each edge having an independent evolutionary rate. And we can visualize the likelihood curve:

```
plot(log(l+1),LL.out,type="l",lwd=3,col="gray",
xlab=expression(paste("Log(",lambda,")")),
        ylab="Log likelihood")
```

This returns figure 2.14.

3. We could easily employ the `sapply( )` function here, but we stick with the "for" loop for transparency.

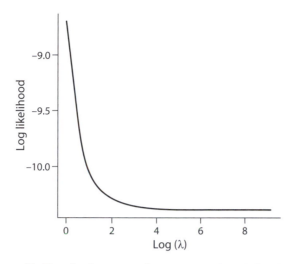

Figure 2.14. Maximum likelihood values across λ parameter values, indicating that a value of 0 is most likely, which corresponds to independent evolutionary rates for each edge.

2.4.2. Edge Length and Rate Transformations

Even if we have a well-supported ultrametric tree, we may wish to alter edge lengths for a number of reasons. First, we may want to change edge distance, say from molecular distances to time. Alternatively, we may wish to alter edge lengths according to some specific model of evolutionary change. For example, we can assess whether early or recent evolutionary change disproportionately explains our ecological pattern of interest by transforming tree edge lengths using Pagel's δ (Pagel 1997, 1999); see chapters 4 and 5 for a worked example. Pagel's δ transforms edge lengths by raising the edge depths to the power of δ. When δ is less than 1, edges near the tips of a tree become shorter, which can be interpreted as evolutionary change being concentrated early in the evolution of the clade. Conversely, δ greater than 1, results in stretching of the edges near the tips, and can be seen as more recent evolutionary change. A δ of 1 is analogous to Brownian motion evolution, and just returns the original tree. We will use the `rescale()` function in the package *geiger* (Harmon et al. 2008). If we look up the help file using `?rescale`, we can see that there are a number of transformations possible. We can see the δ transformation using a random ultrametric tree:

```
library(geiger)
tr<-rcoal(25)
tr.D0.1<-rescale(tr,"delta",0.1)
tr.D1<-rescale(tr,"delta",1)
tr.D10<-rescale(tr,"delta",10)
```

Let's examine how these trees differ visually (fig. 2.15):

```
par(mfrow=c(1,3),cex.main=2)

plot(tr.D0.1,show.tip.label=FALSE, edge.width=2,
    main=expression(paste("A) ",Delta," = 0.1")))
```

A δ = 0.1   **B** δ = 1   **C** δ = 10

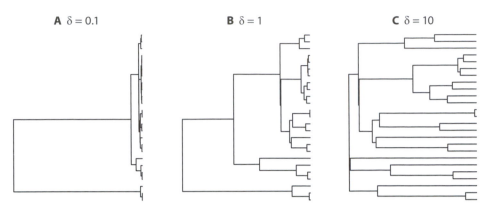

Figure 2.15. Examples of three tree transformations using Pagel's d. δ = 1 returns the original tree and δ < 1 extends internal edges and contracts terminal ones, whereas δ < 1 extends the terminal edges.

```
plot(tr.D1,show.tip.label=FALSE, edge.width=2,
    main=expression(paste("B) ",Delta," = 1")))

plot(tr.D10,show.tip.label=FALSE, edge.width=2,
    main=expression(paste("C) ",Delta," = 10")))
```

Remember that a δ = 1 returns the original tree (fig. 2.15B). The small δ value compresses the terminal edges and stretches the branches deeper in the tree (fig. 2.15A). When we use a large δ value, the terminal edges are stretched, corresponding to high character divergence among closely related taxa (fig. 2.15C).

To assess whether different rates of evolution are associated with branching events (diversification rates), we can transform trees using Pagel's κ (Pagel 1997, 1999). Pagel's κ raises edge lengths to the power of κ. When κ is less than 1, longer edges are shortened more than shorter ones, so that clades with many lineages appear to have more evolutionary change than less diverse ones. A value of κ = 0 sets all branch lengths equal, and is equivalent to a speciational model of evolution. A κ greater than 1 stretches longer edges more, and so clades with fewer lineages undergo relatively more evolutionary change. A value of κ = 1 returns the original tree. Again, we will use the `rescale()` function in *geiger* to do this transformation. Here is the code:

```
tr.K0.1<-rescale(tr,"kappa",0.1)
tr.K1<-rescale(tr,"kappa",1)
tr.K2<-rescale(tr,"kappa",2)

par(mfrow=c(1,3),cex.main=2)
plot(tr.K0.1,show.tip.label=FALSE, edge.width=2,
    main=expression(paste("A) ",Kappa," = 0.1")))
plot(tr.K1,show.tip.label=FALSE, edge.width=2,
    main=expression(paste("B) ",Kappa," = 1")))
plot(tr.K2,show.tip.label=FALSE, edge.width=2,
    main=expression(paste("C) ",Kappa," = 2")))
```

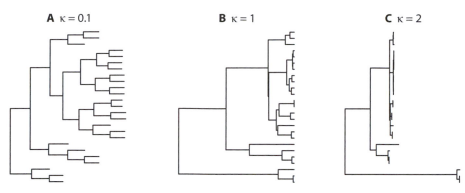

A κ = 0.1   **B** κ = 1   **C** κ = 2

Figure 2.16. Examples of the edge length transformations from different κ values. A value of κ = 1 returns the original tree, κ < 1 results in higher rates of evolution in diverse clades, and κ > 1 suppresses evolution in diverse clades.

This produces figure 2.16. The small κ thus suggests higher rates of evolution within the most diverse clade (fig. 2.16A), and a larger κ suggests higher rates in the less diverse clade (fig. 2.16C). We revisit the κ transformation in chapter 5.

Finally, we may wish to remove the phylogenetic structure of our tree, and Pagel's λ provides a method to do this (Pagel 1999). At the extreme, a complete lack of phylogenetic structure can be represented by a "star" phylogeny (see fig. 2.17 and chapter 3), where all species descend from the root node and are thus all equally related to one another. Pagel's λ transforms our phylogeny by reducing the overall contribution of internal edges relative to the terminal edges. The value λ is scaled from 0 to 1 (it is possible to obtain values of λ greater than one, but these have no clear evolutionary interpretation), with 0 removing all phylogenetic structure and 1 returning our original phylogeny. We can do this by again using *geiger*:

```
tr.L0<-rescale(tr,"lambda",0)
tr.L05<-rescale(tr,"lambda",0.5)
tr.L1<-rescale(tr,"lambda",1)

par(mfrow=c(1,3),cex.main=2)
plot(tr.L0,show.tip.label=FALSE, edge.width=2,
    main=expression(paste("A) ",lambda," = 0")))
plot(tr.L05,show.tip.label=FALSE, edge.width=2,
    main=expression(paste("B) ",lambda," = 0.5")))
plot(tr.L1,show.tip.label=FALSE, edge.width=2,
    main=expression(paste("C) ",lambda," = 1")))
```

And from figure 2.17, we can see clearly that λ = 0 returns a star phylogeny (fig. 2.17A). A λ of 0.5 gives us a phylogeny where the relative importance of the phylogenetic structure is reduced by increasing the amount of species' independent evolution relative to the edges that represent patterns of relatedness (fig. 2.17B). Finally λ = 1 just returns our original phylogeny (fig. 2.17C), which was produced by another run of `rcoal()`.

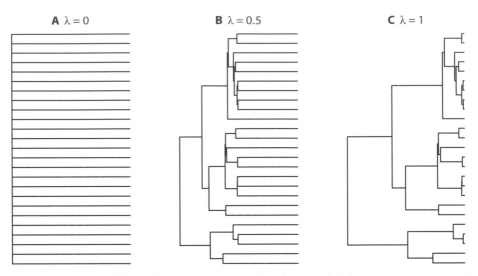

Figure 2.17. Examples of λ transformations showing the reduction of phylogenetic structure. A value of $\lambda = 1$ returns the original tree, and $\lambda < 1$ results in increases in the independent evolution of species (i.e., extending tips) and the expense of internal structure. A value of $\lambda = 0$ removes phylogenetic structure, creating a star phylogeny where all species are equally related to one another.

2.5. CONCLUSION

Phylogenies contain a wealth of information for ecologists, but those using them for eco-phylogenetic analyses often overlook the number of methodological steps and key assumptions that enter into phylogenetic tree reconstruction. We have highlighted a number of approaches for obtaining a phylogenetic tree but our treatment of this topic is certainly not exhaustive. Rather, we present information that will help inform and educate ecologists about phylogeny construction and we encourage the interested reader to explore other sources if they wish to reconstruct their own tree for a particular taxa set.

This chapter also serves at a brief introduction into some of the R packages that we will use to handle and manipulate phylogenetic trees in this book, and we further cover some very standard tree-scaling routines. Again, there are more methods available and our goal for this book is not to train phylogeneticists, but rather to guide ecologists. We hope that knowledge of these tools will help ecologists perform ecophylogenetic analyses with greater confidence.

CHAPTER 3

Phylogenetic Patterns within Communities

Inferring Mechanisms of Ecological Assembly Using Phylogenetic Distances

The history of the past interests us only in so far
as it illuminates the history of the present.
—*Ernest Dimnet*

We live in a nonrandom world. From the elements of the universe and the basic laws of physics emergent order has arisen. One manifestation of this nonrandomness is in the dauntingly complex ecological communities that blanket Earth. Community ecologists are driven to understand patterns of diversity and why certain species coexist, while others are forbidden companions and are unable to persist together. Numerous ecological mechanisms are invoked to explain these patterns; these rely on limitations imposed by the abiotic environment as well as biotic limitations, where certain species exclude others because of exploitative or competitive interactions, or where some species provide additional opportunities for others as a result of beneficial interactions (i.e., facilitation). Classic studies have attempted to explain these patterns of diversity and distribution by measuring the numbers, abundance, and distributions of species (Andrewartha and Birch 1954, MacArthur and Wilson 1967, Kareiva and Levin 2003). Yet more recently, ecologists have come to recognize that these measures have only a limited ability to provide mechanistic insights into the process of community assembly (McGill et al. 2006). One could easily envision visiting two habitats with identical numbers of species; however, in each, the assembly processes has been markedly different. For example, the set of species that occur in high elevation meadows do so because they can tolerate the stressful freezing-thawing and high winds that are commonly found there, mediated through a combination of facilitative and competitive interactions (Lortie et al. 2004, Maestre et al. 2009). Conversely, competition and niche partitioning are likely the most important processes determining which species occur in a low-elevation old field, with the abiotic environment perhaps playing a lesser role, at least at local scales (Tilman 1982, Myster and Pickett 1990).

To truly understand these mechanisms, we need to measure species' differences. While there has been a long history of measuring species' ecological differences as a way to understand where each lives (Raunkiær 1934, Grime 1979) and how they evolved (Harvey and Pagel 1991), the recognition that these differences are also important for understanding species interactions and community assembly has received new attention recently (Walker et al. 1999, Webb 2000, Petchey and Gaston 2002, McGill et al. 2006). Mechanisms influencing patterns of community assembly act on the ecological similarities and differences of organisms, and not on the number of species (Weiher and Keddy 1995b, Diaz and Cabido 2001, Cavender-Bares and Wilczek 2003, Cavender-Bares et al. 2006, McGill et al. 2006,

Cadotte et al. 2011). Phylogenetic patterns of relatedness have provided an especially useful and popular approach to quantifying expected species ecological differences for examining community assembly (Cavender-Bares et al. 2009b, Mouquet et al. 2012, Rolland et al. 2012). As discussed in chapter 1, a phylogenetic approach is predicated on the assumption that the longer two species have evolved independently of one another, the greater the accumulation of ecological differences. This assumption is likely oversimplistic (see chapter 10), but it serves as a good starting point.

Over the past decade, phylogenetic information has been used frequently in community assembly studies to test a basic set of hypotheses about the relatedness of co-occurring species (Mouquet et al. 2012). The principles underlying this usage are deceptively simple. If competitive exclusion is a dominant process and coexistence occurs because species possess sufficient niche differences, then assemblages should be comprised of distantly related species (fig. 3.1). These distantly related species have had sufficient time to accumulate broad differences in their phenotypes and resource requirements, resulting in reduced competition for limited resources. Competitive coexistence is the product of a number of factors, but ultimately requires incomplete resource or niche overlap (Chesson 2000, Adler et al. 2007). The only way that species with identical niches can persist together is if they have identical fitness (e.g., neutrality) (Hubbell 2001, Chave 2004, Adler et al. 2007), which is a rather restrictive condition. As soon as species exhibit a fitness difference, there needs to be some niche difference for coexistence. Thus, the greatest probability of coexistence results from species combinations that have minimal niche overlap.

However, there are two issues that complicate the simplicity of this hypothesis. The first problem is that to date there has not been a thorough assessment of how niche and fitness differences actually evolve (see chapter 10), and only a limited assessment of how

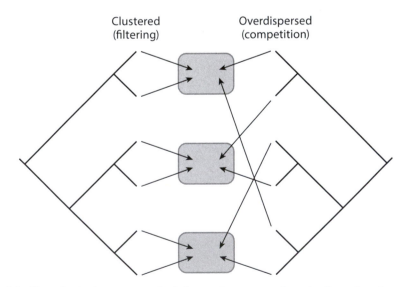

Figure 3.1. Hypothesized patterns of phylogenetic over- and underdispersion in communities. The *gray arrows* indicate which taxa are present in the habitats (*gray squares*). Adapted from Cavender-Bares and Wilczek (2003).

competition varies with phylogenetic distance (Bennett et al. 2013); also, evolutionary models have not yet been used to generate expectations. Thus, we do not know how a phylogeny might represent niche overlap and competition. For example, it is possible to look too deep in the phylogeny, where evolutionary distance becomes so great it no longer correlates with the niche difference required for coexistence.

The second problem with using phylogenies for studies of overdispersion is that if close relatives have very similar fitness values, then they only require small niche differences to coexist. The result of this would be that close relatives might also be able to coexist, and while competition might still drive coexistence, the pattern would not be consistent with figure 3.1 (Narwani et al. 2013, Godoy et al. 2014). These types of concerns about the assumption-laden links between competition and phylogenetic distances have led some to question the utility of using phylogenies to examine community assembly (Gerhold et al. 2015). We would argue that the preponderance of phylogenetically nonrandom communities requires an explanation and more work to link competitive interactions with phylogenetic distances (see chapter 10).

Assemblages containing many closely related species are thought to reflect mechanisms selecting for similar species (fig. 3.1), especially when more distantly related species could potentially disperse into the habitat. Within the current paradigm, researchers often see such a pattern as evidence for environmental filtering, where some aspect of the abiotic environment, such as nutrient limitation, temperature or moisture extremes, or substrate conditions, selects for species with a specific suite of traits. This suite of traits is more likely to be shared among closely related species. For example, plants in the mustard green or crucifer family (Brassicaceae) are often characterized as fast growing and short lived and are often associated with ephemeral or recently disturbed habitats. Thus we might expect clusters of closely related crucifers in frequently disturbed habitats. Similarly, stressful habitats or anthropogenic changes can serve as a strong environmental filter. For example, zooplankton in lakes are more closely related following human-caused disturbance (Helmus et al. 2010).

However, it is likely that other processes could also produce clustered communities. For example, symbiotic relationships can influence the phylogenetic relationships within a community. Mutualistic associations, such as plant-pollinator interactions or mycorrhizal fungi growing with plants, could result in multiple, closely related species being supported by a single mutualist (Bastolla et al. 2009). Similarly, pathogens and other enemies that are shared among closely related species might result in the elimination of entire host clades, and, conversely, resistance to natural enemies might also be phylogenetically patterned. The resulting assemblages may then be comprised of groups of relatively closely related species not affected by the enemy.

An additional challenge to inferring mechanism from a pattern of phylogenetic clustering is the fact that competition could also potentially create clustered assemblages (Mayfield and Levine 2010). If there is a phylogenetically conserved trait that confers a competitive advantage, then a community may be comprised of closely related species because they outcompete other, more distantly related species (fig. 3.2). As stated previously, species with similar fitness need only small niche differences to coexist, and a clade of highly competitive species can therefore coexist because they are relatively similar (Mayfield and Levine 2010).

The complexities outlined above focus our attention on our inferences, essentially linking patterns to specific mechanisms. However, as we have already suggested, the ubiquity

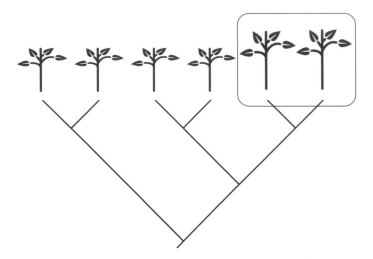

Figure 3.2. Competitive exclusion driving phylogenetic clustering. In this example, the species in the far right clade are taller and thus outcompete other species. Adapted from Mayfield and Levine (2010).

of nonrandomly structured communities deserves investigation. Communities are comprised of organisms that carry with them their unique evolutionary histories, and these histories influence where species can live and how they make their livings today. The question we ask here is *how* to use phylogenetic information to examine patterns within communities (we will examine patterns among communities—phylobetadiversity—in chapter 6). There are a number of different metrics employing phylogenetic information, but they can be broadly categorized into (1) those that examine patterns of relatedness by measuring phylogenetic distances among species (i.e., summing or averaging edge lengths directly), and (2) those that provide a phylogenetic index of diversity (hereon referred to as phylodiversity[1]) within an assemblage by accounting for the distribution of or evenness in edge lengths. The latter often take a more information theoretic approach. The first approach is frequently used to assess mechanisms of assembly and coexistence of species. The second approach often works to incorporate phylogenetic information into more traditional diversity measures to assess the distribution of diversity, its implications for ecosystem function, and to establish conservation prioritization. The applications of phylodiversity metrics are discussed further in chapter 9, here we describe how they are calculated.

3.1. PHYLOGENETIC DISTANCES AND COMMUNITY ASSEMBLY

Looking at figure 3.1, it is obvious that some species are more closely related to one another than to others, and this is straightforward to quantify. Assuming that the phylogeny has meaningful edge lengths, then each edge has a specified distance between nodes (see chapter 2

1. We use "phylodiversity" instead of "phylogenetic diversity" to avoid confusion with a commonly used measured called phylogenetic diversity, and often denoted by PD.

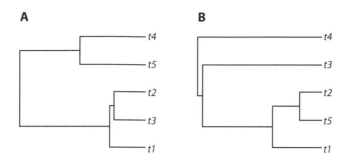

Figure 3.3. Two sample phylogenies produced by `rcoal()` with five tips.

for phylogenetic tree construction and terminology). In R, the phylogenetic tree object has these edge lengths assigned to a unique edge identifier. Let us look at an example using a random tree (remember the values will change every time a random tree is generated):

```
tr<-rcoal(5)
```

Here the argument "5" specifies that this tree will have five tips (species), as shown in figure 3.3A. We use `rcoal()` because it returns an ultrametric (coalescent) tree, whereas the alternative `rtree()` returns a nonultrametric tree. As we have seen already, we can return these edge lengths for a tree in R by calling:

```
tr$edge.length
```

One could go through these, knowing which values correspond to which species and add the appropriate ones together to get the species pairwise distances. There is a function to do this called `cophenetic()`, which we introduced in chapter 2, and to which we will return later on.

One of the ecophylogenetic measures with the longest history is Faith's phylogenetic diversity or Faith's PD (hereafter referred to as PD), first described by Dan Faith (Faith 1992a). There is some confusion about how to employ this measure when moving from a regional species pool to local habitats, with some studies keeping the root node of the regional phylogeny in any assemblage subtree, even if the taxa do not traverse the root; other studies calculate PD using only the assemblage taxa. Within a single assemblage we will define PD as the total amount of evolutionary history represented by a group of extant species excluding the root. To calculate it, we simply sum all the edge lengths from the vector of edge lengths as

```
sum(tr$edge.length)
```

For our assemblage, from the example above, we get a value of 4.648. The PD varies depending on the distribution of node heights within an ultrametric tree. A star phylogeny (e.g., fig. 3.4), where all species are equally related and originate from the root node, has the greatest PD value. If we were to scale the edges (see chapter 2) so that the maximal, or root to tip value is 1, then PD equals species richness (Tucker and Cadotte 2013).

This simple summation works for a single community with a single phylogeny. Most often we have data from multiple sites or plots and would need to calculate PD across all these assemblages. Using a community data matrix (see table 4.1), we can use the `pd()`

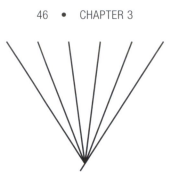

Figure 3.4. An example of a "star" phylogeny, where all taxa are equally related.

function in *picante*. Here we use the *phylocom* data set from *picante* that we introduced in chapter 2; as a reminder, the *phylocom* data object is a list comprised of three components: a phylogeny (`phylocom$phylo`), a community data matrix (`phylocom$sample`), and a trait matrix (`phylocom$traits`). The `pd()` function has three arguments specifying (1) the data matrix, (2) the phylogeny, and (3) whether the calculation of PD should include the root.

Here we will calculate PD for the five communities:

```
library(picante)
data(phylocom)

pd(phylocom$sample,phylocom$phylo,include.root=FALSE)
```

This returns:

	PD	SR
clump1	14	8
clump2a	16	8
clump2b	18	8
clump4	22	8
even	30	8
random	27	8

Here, each site has eight species, but they vary in their phylogenetic diversity because of different patterns of relatedness. The community called "even" has the greatest PD, with the clumped sites having low PD values, because these sites contain close relatives (fig. 3.5). Notice that we included the argument `include.root=FALSE`, which excludes the root node for assemblages that do not include taxa that traverse it. Including the root node has value for relative conservation valuation, which we will discuss in chapter 9.

Differences in PD can point to potential ecological or evolutionary processes that influence species coexistence. To test assembly mechanisms, we will look at the information provided by pairwise distances. Even though we may see differences in PD, we may not get a good sense of how species within an assemblage are related to one another. For example, do we get groups of closely related species that span the entire tree? Or are species somewhat distantly related throughout the tree? Depending on the tree topologies, we could get similar PD values for very different patterns of relatedness; pairwise distances may provide more insight.

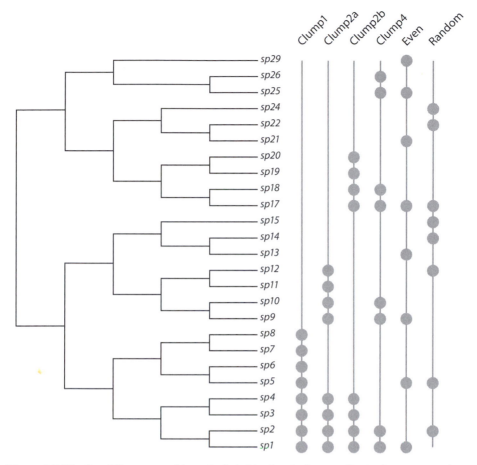

Figure 3.5. The five different assemblages included in the `phylocom` data in the *picante* package.

We can get the pairwise distance matrix using `cophenetic(tr)` for the tree in figure 3.3A, and we get the following distance matrix:

```
          t1          t3          t2          t5          t4
t1  0.0000000   0.8808809   0.8808809   3.023115    3.023115
t3  0.8808809   0.0000000   0.7805563   3.023115    3.023115
t2  0.8808809   0.7805563   0.0000000   3.023115    3.023115
t5  3.0231153   3.0231153   3.0231153   0.000000    1.587386
t4  3.0231153   3.0231153   3.0231153   1.587386    0.000000
```

Here the phylogenetic distance connecting species *t2* and *t4* is 3.023. Compare this to *t2*'s closer relative, species *t3*, for which the pairwise distance is 0.7806. This distance matrix is the basis for many of the phylogenetic measures calculated here.

There are two commonly used measures that we can derive from the pairwise distance matrix: mean pairwise distance (also referred to as MPD) and mean nearest taxon distance (MNTD). The latter, MNTD, is also often referred to as MNND—mean nearest neighbor distance (Webb 2000, Webb et al. 2002). The MPD is simply calculated as the mean of the nondiagonal elements in the pairwise distance matrix. In effect, we can calculate this by taking the lower triangle of a distance matrix. From our previous example with the random tree, we can do this:

```
dist.tr<-cophenetic(tr)
dist.tr<-dist.tr[lower.tri(dist.tr,diag=FALSE)]
```

Here we created a distance matrix object *dist.tr* using the `cophenetic()` function, and then we subset that matrix using the `lower.tri()` function, with the argument `diag=FALSE` excluding the diagonal elements (i.e., the zeros). To calculate MPD, we just take the mean of these values:

```
mean(dist.tr)
```

This returns a value of 2.22684. We can use the `mpd()` function in *picante*, but to do this we need a community matrix. For a single assemblage, this can be done by creating a community in which all species are present with the following:

```
a<-matrix(c(1,1,1,1,1),nrow=1,
        dimnames=list("s1",c("t1","t2","t3","t4","t5")))
mpd(a,cophenetic(tr))
```

The `mpd()` function here returns the identical MPD value of 2.22684.

The MNTD can also be calculated from the distance matrix, except this time we are looking for the smallest nondiagonal value for each species. There are a number of ways to do this, but here we will use a simple `apply()` function (which is essentially a function that repeats some other function or action across data) with our *tr* phylogeny object created above. We need to first get rid of the diagonal zeros, since these are technically the minimum values in the distance matrix (the *picante* function `mntd()` that we introduce below automatically gets rid of the zeros), by turning them into NAs, which are then excluded in many functions.

```
b<-cophenetic(tr)
diag(b)<-NA
apply(b,margin = 2,min,na.rm=TRUE)
```

This returns the minimum distances for each species in the assemblage, and we need the mean value:

```
mean(apply(b,2,min,na.rm=TRUE))
```

This returns 1.12343. When we use the `mntd()` function from *picante*, again using the community matrix, we get the same value:

```
mntd(a,cophenetic(tr))
```

The real benefit of the `pd()`, `mpd()`, and `mntd()` functions in *picante* is that they can calculate their respective measures across multiple sites. Depending on the nature of those sites, and the questions of interest, further statistical analyses can be carried out on these PD, MPD, or MNTD values. Here we will use the Jasper Ridge plant community data to examine these metrics. We need to load both the Jasper Ridge community matrix (jasper_data.csv), which we will call *j.com,* and the phylogeny for these species (jasper_tree. phy), which we will call *j.tree.*

```
j.tree<-read.tree("…/jasper_tree.phy")
j.com<-read.csv("…/jasper_data.csv",row.names=1)
```

With this data we can use the pd() function and assign it to an object name, *pd.out*.

```
pd.out<-pd(j.com,j.tree,include.root=FALSE)
```

To examine the results without printing out the whole data frame on screen, we can use the head() function to see the first six rows.

```
head(pd.out)
```

This returns:

```
          PD    SR
J11   831.4887  13
J110  501.8703   9
J12   769.9249  14
J13   712.0532  13
J14   640.5896  11
J15   640.2884  13
```

Here we can see a range of PD and richness values, and interestingly, there are a number of 13-species plots that vary in their PD values (one with 831 million years of evolution and another with just 640 million). Next let's plot their relationship (fig. 3.6):

```
plot(pd.out$SR,pd.out$PD,
xlab="Species richness",ylab = "PD",pch=16)
```

Similarly, we can examine how MPD and MNTD are related to richness, but mpd() and mntd() return vectors, and not a data frame as with pd(). We can simply add the vectors

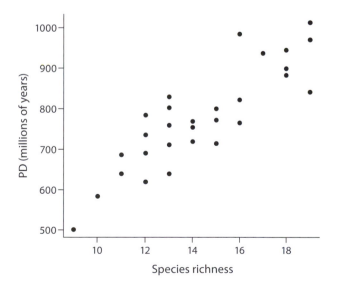

Figure 3.6. The relationship between species richness and PD for the Jasper Ridge plant community data.

as columns to the object *pd.out* from above. We can add MPD and MNTD as new columns because the input data is in the same order (but there are ways to ensure data are ordered the same by using the `match()` function). Be careful if you sort the community matrix between calls to the functions, because then the values may not be in the correct order.

```
pd.out[,3]<-mpd(j.com,cophenetic(j.tree))
pd.out[,4]<-mntd(j.com,cophenetic(j.tree))
```

Let's then make the column names meaningful:

```
names(pd.out)[3:4]<-c("MPD","MNTD")
```

Now we can plot these as multiple panels within the same figure (fig. 3.7).

```
par(mfrow=c(1,2))

plot(pd.out$SR,pd.out$MPD,xlab="Species richness",
     ylab = "MPD (millions of years)",pch=16)

plot(pd.out$SR,pd.out$MNTD,xlab="Species richness",
     ylab = "MNTD (millions of years)",pch=16)
```

Interpreting PD, MPD, or MNTD can be complicated if sites have different species richness. The PD is usually positively related with richness (fig. 3.8A), which is another reason why researchers often prefer to examine MPD and MNTD. However, there are subtle influences of richness on MPD and MNTD, especially moving from low richness communities to those that are species rich (Clarke and Warwick 1998, Cadotte et al. 2010a). Mean MPD for randomly assembled communities remains constant across a richness gradient, but the variance declines (fig. 3.8B). This would mean that if we had few species-poor assemblages, their values could be extremely large or small just due to the effect of randomly sampling a small proportion of the edges of the phylogenetic tree. Mean and variance of MNTD decrease with increasing richness (fig. 3.8C).

If we are interested in the assembly of communities, then we might like to know whether assemblages consist of closely (clustered) or distantly (overdispersed) related species. Of course the answer to this question only makes sense if we ask what the relatedness is relative

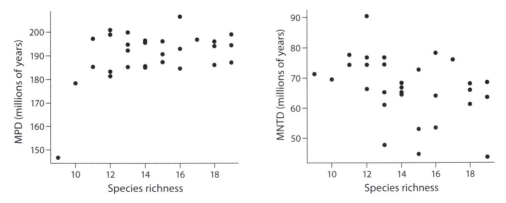

Figure 3.7. The relationships between species richness and MPD and MNTD for the Jasper Ridge plant community data.

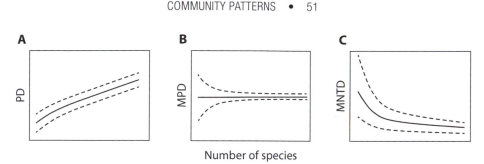

Figure 3.8. The behavior of PD, MPD, and MNTD across species richness gradients. The *solid lines* indicate mean values and *dashed lines* show the 95% confidence interval.

to. In order to test whether communities are phylogenetically clustered or overdispersed, we can compare actual PD, MPD, and MNTD values to those expected by chance. To do this we use a null model where we randomize the assemblages or the evolutionary relationships among species. There are a number of ways to create and use null assemblages, and we explore this further in chapter 4, but here we will use one of the simplest null models, randomly shuffling the tip labels on the phylogeny. Thus when we recalculate our metrics for the communities, the phylogenetic distances will not correspond to the actual distances separating species, but the overall structure of the phylogenetic topology is retained. The random shuffling of tip labels is done with this:

```
sample(tr$tip.label,length(tr$tip.label), replace=FALSE)
```

We can then put this in a `for()` loop or `apply()` function (the specific form of the `apply()` function to do this is the function `replicate()`) and create a relabeled phylogeny and calculate our metric, say, 5,000 times. We can then standardize the metrics by calculating an effect size against the null values using a *z*-value (Webb et al. 2002). The *z*-value is a dimensionless value that indicates the number of standard deviations an observation is from the mean of the distribution. It is calculated as

$$z = \frac{x-u}{\sigma} \tag{3.1}$$

where x is the observation (for example, the MPD value for a community), μ is the mean, and σ is the standard deviation of the null or random distribution. A positive value corresponds to the number of standard deviations above the mean and a negative value to the number of standard deviations below the mean. Thus, the standardized value of PD is

$$z_{PD} = \frac{PD_{obs} - \mu_{nullPD}}{\sigma_{nullPD}} \tag{3.2}$$

where μ and σ are calculated from randomized values. The calculation for standardized MPD and MNTD values is identical:

$$z_{MPD} = \frac{MPD_{obs} - \mu_{nullMPD}}{\sigma_{nullMPD}}; z_{MNTD} = \frac{MNTD_{obs} - \mu_{nullMNTD}}{\sigma_{nullMNTD}} \tag{3.3}$$

These last two standardized measures are similar to those used by Webb et al. (2002) as *NRI* (net relatedness index) and *NTI* (nearest taxon index), respectively, except they represent the negative of the *z*-values: $NRI = -z_{mpd}$ and $NTI = -z_{mntd}$. Negative *NRI* and *NTI* values indicate overdispersion and positive values clustering. *Picante* calculates the *z*-values and they should be multiplied by −1 to turn them into *NRI* and *NTI* values.

The functions to standardize PD, MPD, and MNTD in *picante* (ses.pd(), ses.mpd(), and ses.mntd()) are fairly straightforward and return a wealth of information (note that the package *pez* also calculates these). Using the *phylocom* example data in *picante* we can look at how the different assemblages in figure 3.5 are reflected in the standardized MPD and MNTD values:

```
ses.mpd(phylocom$sample,cophenetic(phylocom$phylo),runs=999)
```

The output is shown in table 3.1a and reports the number of species, the observed MPD value, and the mean and standard deviation of MPD from the null randomization. It also reports the observed rank, which is where the observed value is located when the null values are ordered from smallest to largest. Thus, a rank of 1 means that the observed MPD value is smaller than all the null values. The *z*-value is reported, the sign indicates clustering (negative) or overdispersion (positive), and the magnitude is the number of standard deviations the observation is from the mean. The larger the absolute value, the more likely it is that the observed value is significantly different than the null expectation. The reported *P*-value is calculated as: $P = mpd.obs.rank / (runs + 1)$, and is a two-tailed test; thus, significance at the threshold $\alpha = 0.05$ level is achieved when $P \leq 0.025$ or $P \geq 0.975$. Table 3.1b reports identical results for MNTD. Given the significance criterion, the clumped and even assemblages are significantly different than the null expectation. The only difference between MPD and MNTD is "clump4," which

TABLE 3.1.
Results from *ses.mpd* (a) and *ses.mntd* (b) functions in *picante*.

a) MPD

Plot	ntaxa	mpd. obs	mpd.rand. mean	mpd. rand.sd	mpd.obs. rank	mpd. obs.z	mpd. obs.p	runs
clump1	8	4.857	8.319	0.323	1	−10.728	0.001	999
clump2a	8	6.000	8.331	0.305	1	−7.641	0.001	999
clump2b	8	7.143	8.321	0.324	7	−3.635	0.007	999
clump4	8	8.286	8.328	0.319	370.5	−0.132	0.371	999
even	8	8.857	8.316	0.340	996.5	1.590	0.997	999
random	8	8.429	8.311	0.336	567.5	0.350	0.568	999

b) MNTD

Plot	ntaxa	mntd. obs	mntd.rand. mean	mntd. rand.sd	mntd.obs. rank	mntd. obs.z	mntd. obs.p	runs
clump1	8	2	4.703	0.665	1	−4.065	0.001	999
clump2a	8	2	4.714	0.643	1	−4.225	0.001	999
clump2b	8	2	4.738	0.643	1	−4.258	0.001	999
clump4	8	2	4.696	0.645	1	−4.181	0.001	999
even	8	6	4.708	0.630	998	2.051	0.998	999
random	8	5	4.760	0.635	616	0.378	0.616	999

includes species sister-pairs dispersed through the phylogeny (fig. 3.5). This assemblage has small MNTD values because each species has a close relative in the assemblage, but the MPD is no different from the null expectation because the distance matrix contains both very large and very small distances and the average of these is roughly equal to the overall average edge length. Thus, comparing MNTD and MPD can be valuable, and can tell us something about the phylogenetic depth that best corresponds to nonrandom patterns (Swenson 2011, Swenson 2014). In general, MNTD tells us more about patterns near the tips whereas MPD is more influenced by phylogenetic distances deeper in the phylogeny.

The assemblages in this example's data are selected to maximize clustering and overdispersion, and the large z-values reflect this. With real data, interpreting the values may require a little more subtlety. Here we will go through the `ses.mpd()` results using the Jasper Ridge plot data example. We will store the output in an object called *smpd.out*.

```
smpd.out<-ses.mpd(j.com,cophenetic(j.tree),runs=999)
```

According to the output (table 3.2), only plot J110 is significantly clustered ($P < 0.025$) and only J11 is overdispersed ($P > 0.975$). These plots could be unique in some way (e.g., recent disturbance, higher elevation, etc.), but we normally expect that one in twenty will have a significant value because of random chance. Two significant findings out of thirty does not instill confidence in a general conclusion. Still, we could use the information in *smpd.out* to perform additional analyses. Let's create a hypothetical treatment for these plots, with half the plots on a dry slope and half on a moist slope. To make this interesting, let's assign the plots to these treatments nonrandomly, with negative z-values belonging to "dry" sites and positive z-values to "moist" sites. We can add a column to *smpd.out* for these treatment designations, and make them a factor (the last line of code below).

```
smpd.out$treat<-NA
smpd.out$treat[smpd.out$mpd.obs.z<0]<-"Dry"
smpd.out$treat[smpd.out$mpd.obs.z>0]<-"Wet"
smpd.out$treat<-factor(smpd.out$treat)
```

Now that the plots belong to two treatments, we can then run standard statistical analyses (e.g., analysis of variance using the function `aov()`) to see if they are significantly different from one another:

```
mod<-aov(smpd.out$mpd.obs.z~smpd.out$treat)
summary(mod)

                 Df   Sum Sq   Mean Sq   F value      Pr(>F)
smpd.out$treat    1   17.7419  17.7419    34.301   2.693e-06 ***
Residuals        28   14.4829   0.5172

---
Signif. codes:  0 '***' 0.001 '**' 0.01 '*' 0.05 '.' 0.1 ' ' 1
```

The two treatments are highly significantly different (shown in fig. 3.9). This hypothetical example illustrates that even though individual communities may not be significantly

TABLE 3.2.
Results from *ses.mpd* for the Jasper Ridge plots.

Plot	ntaxa	mpd.obs	mpd.rand. mean	mpd. rand.sd	mpd.obs. rank	mpd.obs.z	mpd.obs.p	runs
J11	13	199.466	195.360	8.767	639	0.468	0.639	999
J110	9	146.623	194.986	12.668	5	−3.818	0.005	999
J12	14	195.501	195.171	8.311	465	0.040	0.465	999
J13	13	192.153	195.471	8.875	295	−0.374	0.295	999
J14	11	185.207	195.447	10.678	151	−0.959	0.151	999
J15	13	185.166	195.425	8.833	126	−1.161	0.126	999
J16	16	192.885	195.212	7.170	312	−0.325	0.312	999
J17	12	183.142	195.353	9.521	121	−1.283	0.121	999
J18	12	200.507	195.282	9.376	673	0.557	0.673	999
J19	14	185.401	195.763	7.765	100	−1.334	0.100	999
J21	10	178.371	195.624	11.105	77	−1.554	0.077	999
J210	12	181.322	195.764	9.186	65	−1.572	0.065	999
J22	13	194.583	195.599	8.512	378	−0.119	0.378	999
J23	18	193.912	195.572	5.703	357	−0.291	0.357	999
J24	18	195.734	195.486	5.877	451	0.042	0.451	999
J25	11	197.115	195.675	9.710	502	0.148	0.502	999
J26	16	184.387	195.028	6.734	75	−1.580	0.075	999
J27	16	206.346	195.492	6.719	976	1.615	0.976	999
J28	15	187.270	195.032	7.374	144	−1.053	0.144	999
J29	12	198.858	195.258	9.602	598	0.375	0.598	999
J31	13	199.677	195.068	8.615	683	0.535	0.683	999
J310	18	185.820	195.231	5.901	61	−1.595	0.061	999
J32	14	196.488	195.631	7.806	506	0.110	0.506	999
J33	14	184.982	195.416	7.640	94	−1.366	0.094	999
J34	19	186.886	195.366	5.368	77	−1.580	0.077	999
J35	15	195.971	195.318	7.218	478	0.090	0.478	999
J36	15	190.610	195.381	7.423	231	−0.643	0.231	999
J37	17	196.799	195.004	6.133	598	0.293	0.598	999
J38	19	194.256	195.442	5.336	367	−0.222	0.367	999
J39	19	198.711	195.181	5.209	749	0.678	0.749	999

clustered or overdispersed, there may be underlying gradients or conditions that influence their diversity (although in this case we created the diversity gradient ourselves). This is often the case when researchers examine the influence of elevation on community phylodiversity and observe that high elevation sites have low phylodiversity compared to low elevation sites (e.g., Cadotte et al. 2013).

When using functions like `ses.mpd()`, it is important to note that individual null values are not stored. If they are needed for further analysis (e.g., producing a histogram

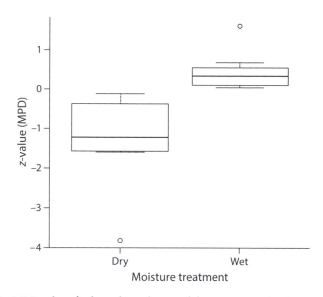

Figure 3.9. The MPD values for hypothetical wet and dry treatments for the Jasper Ridge plots.

of values), the `replicate()` and `sample()` functions outlined previously are required, and this will be covered in the next chapter.

3.2. CALCULATING COMMUNITY DIVERSITY METRICS

There are a multitude of ways to quantify community taxonomic diversity (Magurran 2003) and even more ways to quantify phylodiversity (e.g., Faith 1992a; Clarke and Warwick 1998; Barker 2002; Pavoine et al. 2005; Hardy and Senterre 2007; Helmus, Bland, et al. 2007; Schweiger et al. 2008; Allen et al. 2009; Cadotte, Davies, et al. 2010; Mouchet and Mouillot 2011; Scheiner 2012). In chapter 9 we will explicitly deal with measures that quantify individual species evolutionary distinctiveness (ED) and their conservation evaluation (Redding and Mooers 2006, Isaac et al. 2007, Redding et al. 2008, Rosauer et al. 2009, Cadotte and Davies 2010), but we also include several ED-based phylodiversity measures here. Our working definition of ED is the amount of unique phylogenetic edge length represented by a species in a phylogenetic tree (Redding and Mooers 2006a, Isaac et al. 2007, Redding et al. 2008).

From table 3.3, it is clear that some metrics are quite similar to one another, with many metrics using pairwise distance measures and species abundances. Some of the measures are readily available in R (e.g., in *picante* and *pez*), others are easily calculated, and others still are not available in an R package and are nontrivial to code. A few metrics have problematic assumptions or qualities that limit their usefulness. We will examine several metrics from table 3.3 in the following sections.

The purpose of using phylodiversity is to capture additional information contained in phylogenies, beyond species richness and abundance. PD, MPD, and MNTD, which we are now familiar with, calculate diversity at the site level and reflect certain aspects of the tree or pairwise distance matrix, but do not directly capture subtle topological patterns, such as the

TABLE 3.3.
Different measures of community phylodiversity.

Metric	Equation	Definition	In R?	Reference
Based on total edge lengths or nodes				
PD (phylogenetic diversity)	$\sum_{e \in z(T)} \lambda_e$	Sum of total edge lengths connecting species together.	*picante* *pez*	(Faith 1992a)
PD$_{ave}$ (Average PD)	$\dfrac{PD}{S}$	PD divided by the number of species in the assemblage.	NA	(Clarke and Warwick 2001)
PD$_{ab}$ (Abundance weighted PD)	$\sum_{e \in z(T)} \left(\lambda_e \cdot \sum_{k}^{S_e} \dfrac{n_i}{N} \right)$	Sum of edge lengths, where edges are scaled by proportional abundances of subtending species.	NA	(Barker 2002)
PAE (Phylogenetic abundance evenness)	$\dfrac{PD + \sum_{i=1}^{S} \lambda_i \cdot (n_i - 1)}{PD + (\bar{n} - 1)\sum_{i=1}^{S} \lambda_i}$	Measures the distribution of abundances across the phylogeny relative to an even distribution.	*pez*	(Cadotte, Davies, et al. 2010)
IAC (Imbalance of abundances at the clade level)	$\dfrac{\sum_{i=1}^{S} |n_i - \hat{n}_i|}{\nu}$	The deviation of species' abundances from a null expectation, where abundance is split according to the number of shared lineages.	*pez*	(Cadotte, Davies, et al. 2010)
Based on pairwise distances				
MPD (mean pairwise distance)	$\dfrac{1}{D} \sum_{i} \sum_{j \neq 1} d_{ij}$	Mean distances from pairwise distance matrix.	*picante, pez*	(Webb et al. 2002, Kembel et al. 2010)
MNTD (mean neighbor taxon distance)	$\dfrac{1}{S} \sum_{i} d_{i\,min}$	Mean shortest distance for each species from the distance matrix.	*picante, pez*	(Webb et al. 2002, Kembel et al. 2010)
D$_D$ (Pure diversity)	$\sum_{i} d_{i\,min}$	Sum of nearest neighbor distances.	NA	(Faith 1994, Solow and Polasky 1994)
MPD$_{ab}$ (Abundance weighted MPD)	$\dfrac{1}{D} \sum_{i} \sum_{j \neq 1} \left(d_{ij} \cdot \dfrac{n_i \cdot n_j}{\sum_{i} \sum_{j \neq 1} n_i \cdot n_j} \right)$	Mean pairwise distances, adjusted by the proportional product of the abundances of the species making that distance.	*picante, pez*	(Webb et al. 2002, Kembel et al. 2010)

Metric	Equation	Definition	In R?	Reference
MNTD$_{ab}$ (Abundance weighted MNTD)	$$\frac{1}{S}\sum_i\left(d_{i\min}\cdot\frac{n_i\cdot n_{j\min}}{\sum_i\sum_{j\neq 1}n_i\cdot n_{j\min}}\right)$$	Mean shortest distances, adjusted by the proportional product of the abundances of the species making that distance.	*picante, pez*	(Webb et al. 2002, Kembel et al. 2010)
J (Intensive quadratic entropy)	$$\frac{\sum_i\sum_{j\neq 1}d_{ij}}{S^2}$$	Average distance between two randomly chosen species.	NA	(Izsák and Papp 2000)
F (Extensive quadratic entropy)	$$\sum_i\sum_{j\neq 1}d_{ij}$$	Sum of all pairwise distances.	NA	(Izsák and Papp 2000)
QE (Rao's)	$$\sum_i\sum_{j\neq 1}(d_{ij}\cdot p_i p_j)$$	Simpson's type diversity index, where the product of species frequencies is weighted by phylogenetic distances.	*ade4 picante, pez*	(Clarke and Warwick 1998, Pavoine et al. 2005, Hardy and Senterre 2007)
H$_d$	$$H_d=-\sum_i^S p_i\ln\left(1-\sum_{j\neq 1}d_{ij}p_j\right)$$	Entropic measure of the distribution of abundance weighted pairwise distances.	*pez*	(Allen et al. 2009)
AvTD (Average taxonomic distinctiveness) *Analogous to MPD	$$\frac{\sum_i\sum_{j\neq 1}d_{ij}}{s(s-1)/2}$$	Average path length connecting species in an assemblage.	NA	(Clarke and Warwick 1998)
AvTDab (Abundance weighted average taxonomic distinctiveness) *Analogous to MPDab	$$\frac{\sum_i\sum_{j\neq 1}(d_{ij}\cdot n_i n_j)}{\sum_i\sum_{j\neq 1}n_i n_j}$$	Gives the expected edge length connecting to randomly selected species in an assemblage.	NA	(Clarke and Warwick 1998)

Based on covariance matrix

Metric	Equation	Definition	In R?	Reference
PSV (Phylogenetic species variability)	$1-\overline{c}$	Variability in an unmeasured neutral trait or the relative amount of unshared edge length.	*picante, pez*	(Helmus, Bland, et al. 2007)
PSR (Phylogenetic species richness)	$n\cdot PSV$	The deviation from species richness, penalized by close relatives.	*picante, pez*	(Helmus, Bland, et al. 2007)

(*continued*)

TABLE 3.3. (*Continued*)

Metric	Equation	Definition	In R?	Reference
PSE (Phyloge-netic species evenness)	$$\dfrac{N \cdot \mathrm{diag}(C)'N - N'CN}{n^2 - (n \cdot n)}$$	Abundance weighted PSV.	*picante, pez*	(Helmus, Bland, et al. 2007)
Based on species' evolutionary distinctiveness				
H'$_{ED}$	$$-\sum_{i=1}^{S} \dfrac{ED_i}{PD} \cdot \ln \dfrac{ED_i}{PD}$$	Entropic measure of evolutionary distinc-tiveness values. Anal-ogous to the Shannon Index.	*pez*	(Cadotte, Davies, et al. 2010)
E$_{ED}$	$$\dfrac{H'_{ED}}{\ln S}$$	Equitability of H'$_{ED}$	*pez*	(Cadotte, Davies, et al. 2010)
H'$_{AED}$	$$-\sum_{i=1}^{S} \dfrac{n_i \cdot AED_i}{PD} \cdot \ln \dfrac{n_i \cdot AED_i}{PD}$$	Entropic measure of evolutionary dis-tinctiveness values— weighted by species abundances.	*pez*	(Cadotte, Davies, et al. 2010)
E$_{AED}$	$$\dfrac{H_{AED}}{\ln N}$$	Equitability of H'$_{AED}$	*pez*	(Cadotte, Davies, et al. 2010)
^{q}D(P)	$$\left(\sum_{i=1}^{S} ED_i^q \right)^{1/(1-q)}$$	A modification of the Hill index, weighting a species' proportional abundance by its rela-tive share of phyloge-netic information.	NA	(Scheiner 2012)

*Variables: $e \in z(T)$ denotes the edges, e, in set z for tree T; λ_e is an edge with length λ; λ_i is a terminal edge with length λ corresponding to species i; S_e is the number of species originating from edge e; N is the total abundance of all species; n_i is the abundance of species i; d_{ij} is the pairwise phylogenetic distance connecting two species; $d_{i\min}$ is the minimum pairwise distance connecting species i to any other species in the set; $n_{j\min}$ is the abundance of the species with minimum distance to species i; $\bar{c}$ is the mean of the off-diagonal elements of the phylogenetic covariance matrix, $\mathbf{C}$; p is the propor-tional abundance (n/N); $\hat{n}_i$ is the null expectation that abundances are split evenly at each node split from root to tip, for a fully dichotomous tree, given as $N/2^v$, where v is the number of nodes in the phylogeny; ED is evolutionary distinctiveness, which is the proportion of the total edge length repre-sented by species i (see chapter 9 for exploration of this concept and measure); AED is a measure of abundance weighted evolutionary distinctiveness, again discussed in chapter 9; q is an exponent that determines how proportions are weighted.

**Please note that the equations always correspond to a single assemblage. For many there would be an additional Σ for summing or averaging across sites.

distribution of information in the tree, tree balance, or distinctiveness of lineages. Further, standard metrics of PD, MPD, and MNTD do not include abundance information, which we might want to consider. We will now go through the steps required to calculate a number of metrics, using the Jasper Ridge data, followed by a brief look at their redundancies/complementarities.

Recall that we already have the data frame *pd.out*, which contains PD, MPD, and MNTD values for the Jasper Ridge plots. We will attach the rest of the metric calculations to this data frame. To create a new column, we only need to name it; let's look at this with a measure of the average PD per species that we will call PD_{ave}. Even though PD_{ave} is not in an R package, it is straightforward to calculate:

```
pd.out$PDave <-pd.out$PD / pd.out$SR
```

Now we have a column called "PDave" added to our output table. We will continue this procedure for the other metrics. Next we will calculate PAE and IAC (see table 3.3), both of which use the *pez* package. The functions in *pez* use a combined data type, called a **comparative.comm** object that includes the community matrix with the phylogeny and trait matrix. We use **comparative.comm()** to create this data object, then the phylodiversity functions use this combined data object:

```
library(pez)
j.pez<-comparative.comm(j.tree,as.matrix(j.com))
```

Now that we have the correctly formatted data object, we will use **pez.evenness** to calculate PAE and IAC. It is important to note that **pez.evenness()** actually calculates a number of measures defined as "evenness" measures by Pearse et al. (2014). We calculate them by:

```
out<-pez.evenness(j.pez)
pd.out$PAE<-out$pae
pd.out$IAC<-out$iac
```

Next we will calculate abundance weighted versions of our pairwise distance metrics, MPD_{ab} and $MNTD_{ab}$. These are easy to generate; they rely on the same **mpd()** and **mntd()** functions used earlier with a change in just a single argument, setting **abundance.weighted** to **TRUE**.

```
pd.out$MPDab<-mpd(j.com,cophenetic(j.tree),abundance.
    weighted=TRUE)
```

```
pd.out$MNTDab<-mntd(j.com,cophenetic(j.tree),abundance.
    weighted=TRUE)
```

Two of the measures in table 3.3, AvTD and $AvTD_{ab}$ from Clarke and Warwick (1998), are analogous to MPD and MPD_{ab}, respectively. They are included in table 3.3 to highlight two points. The first is that similar metrics have been created independently by multiple authors (in this case by Clarke and Warwick 1998, and Webb 2000). Second, they are highlighted to give proper credit to the pioneering work by Clarke and Warwick, who were ahead of their time. Next we will calculate pure diversity (D_D). This metric is not currently found in any available package, so we will create a function to calculate it here, which we will call **pureD**. We write it to match the arguments of **mpd()** in order to stay consistent:

```
pureD<-function(samp,dis){
    N <-dim(samp)[1]
    out <-numeric(N)
    for(i in 1:N){
        sppInSample <-names(samp[i,samp[i,] > 0])
        if (length(sppInSample) > 1){
            sample.dis <-dis[sppInSample, sppInSample]
          diag(sample.dis) <-NA
            out[i] <-sum(apply(sample.dis, 2, min, na.rm
                    = TRUE))
            }
            else {
                out[i] <-NA
                }
        }
        out
    }
```

And now we can use this function:

```
pd.out$pureD <-pureD(j.com,cophenetic(j.tree))
```

Let's also create functions for intensive and extensive quadratic entropies (J and F from table 3.3, respectively). We will put these in the same function, with the same arguments as above, but we will add an argument for the type of measure, intensive (`int`) or extensive (`ext`).

```
quad.ent <-function(samp,dis,type = c("int","ext")){
    N <-dim(samp)[1]
    out <-numeric(N)
    for(i in 1:N){
        sppInSample <-names(samp[i,samp[i,] > 0])
        if (length(sppInSample) > 1){
            sample.dis <-dis[sppInSample, sppInSample]
          sample.dis<-sample.dis[lower.tri(sample.dis)]
            if (type == "int"){
                out[i] <-sum(sample.dis)/length(sppInSample)^2
                }
            if (type == 'ext'){
                out[i] <-sum(sample.dis)
                }
            }
            else {
                out[i] <-NA
                }
        }
        out
        }
```

We can then calculate these measures by running our function as follows:

```
pd.out$Intensive <-quad.ent(j.com,cophenetic(j.tree),"int")
pd.out$Extensive <-quad.ent(j.com,cophenetic(j.tree),"ext")
```

Next, let us look at the QE measure. This measure is presented with slight modifications in the literature, and is part of a group of similar metrics based on Simpson's or Rao's diversity (Zoltan 2005, Hardy and Senterre 2007). Rao's diversity is usually presented as:

$$QE = \sum_i \sum_{j \neq 1} (d_{ij} \cdot p_i p_j) \tag{3.4}$$

with d_{ij} being some measure of the distance between species i and j, which in our case is phylogenetic distance. When $d_{ij} = 1$ for all $i \neq j$, this measure is analogous to Simpson's Index (Zoltan 2005):

$$S_D = 1 - \sum_{i=1}^{s} p_i^2$$

where p_i is the proportional abundance of species i. Here, we will use the phylogenetic Rao's D function in *picante*:

```
outD<-raoD(j.com,j.tree)
```

This returns a number of calculations, including Rao's D for each sample, an among sample similarity (see chapter 7), average local diversity, and a suite of beta diversity measures. To pull out the within sample values we can simply call

```
pd.out$Rao<-outD$Dkk
```

Next, we will move on to metrics defined by Helmus, Bland, et al. (2007) based on the covariance matrix. These are a nice set of measures that are scaled between 0 and 1, where 1 is analogous to covariance of 0, which is represented by a star phylogeny. Their three measures represent phylogenetic information in different ways. Phylogenetic species variability (PSV) is a measure of how much species should vary from one another on the phylogeny. As we stated above, species should maximally vary when they are in a star phylogeny; this measure quantifies the relative amount of unshared branch length. The second measure, phylogenetic species richness (PSR), simply multiplies PSV by the number of species. When interpreting this metric, it might be helpful to remember that species richness is just the maximal relatedness possible (i.e., all species equally distantly related) when a phylogenetic tree is scaled with a root to tip distance of 1 (Tucker and Cadotte 2013); thus, the smaller the value of PSR, the more close relatives there are in the phylogeny. Finally, phylogenetic species evenness (PSE) scales PSV by abundances. In essence, this metric creates polytomies below the species level, each individual represented by a single branch of length 0, thus inflating the covariance matrix. These metrics are available in *picante*, and we will now add them to our data frame. Each of the calls to the functions psv(), psr(), and pse() returns a data frame with additional information beyond the metric of interest (like species richness). We only want the column with the metric so notice that we subset the output from the function call using square brackets, [ ], and specify the first column [,1]:

```
pd.out$PSV <-psv(j.com,j.tree)[,1]
pd.out$PSR <-psr(j.com,j.tree)[,1]
pd.out$PSE <-pse(j.com,j.tree)[,1]
```

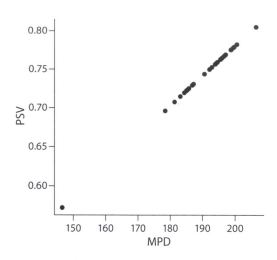

Figure 3.10. The strong correlation between MPD and PSV, highlighting the mathematical linkage between the two.

Superficially, it looks as if these metrics are calculated in a very different way than the previous metrics we have discussed (see equations in table 3.3). However, PSV is actually analogous to MPD, and they just take different routes to the same value. Figure 3.10 shows the relationship between these two values, and the code to show this is:

```
plot(pd.out$MPD,pd.out$PSV,xlab="MPD",ylab="PSV",pch=16)
```

Finally, we will calculate the diversity measures based on ED values. First we will start with the final metric in table 3.3, the phylogenetic modification of the Hill number by Scheiner (2012). The Hill number was originally created by Mark Hill (Hill 1973) as a measure of species diversity, and is known to be generalizable to other diversity measures (Heip et al. 1998). This metric does not currently exist in R, so we will need to create one, based on the `evol.distinct()` function in *picante*, and we name it dDP:

```
dDP <-function(phy,com,q=2){
  ed.list <-evol.distinct(phy)
  out<-numeric(nrow(com))

for (i in 1:nrow(com)){
    ed.tmp<
-ed.list[match(colnames(j.com[i,j.com[i,]>0]),
as.character(ed.list$Species)),]
    l<-ed.tmp$w/sum(ed.tmp$w)
    out[i]<-sum(l^q)^(1/(1-q))

  }
out
        }
```

Now let's add it to our data frame:

```
pd.out$dDP <- dDP(j.tree,j.com, q = 2)
```

One issue with using this metric is in choosing what value of q to use. Small q values (e.g., → 1) give higher weight to rare species and larger values (e.g., → 3) give higher weight to common species (Leinster and Cobbold 2011). Different values of q correspond to different diversity metrics (e.g., q = 2 is equivalent to an exponential Shannon) and so have different interpretations (Leinster and Cobbold 2011). For our example we used q = 2 for simplicity.

Finally, we will calculate the entropic ED measures described by Cadotte, Davies, et al. (2010). These measures are analogous to the Shannon index and have corresponding equitability measures of evenness. In the unweighted versions (H_{ED} and E_{ED}) we can use the `pez.shape()` functions from the package *pez*:

```
out.s<-pez.shape(j.pez)
pd.out$Hed <-out.s$Hed
pd.out$Eed <-out.s$Eed
```

We then use the equivalent abundance weighted versions. We actually already calculated these when we previously ran the `pez.evenness()` function for PAE and IAC, so we can simply pull out the relevant values:

```
pd.out$Haed <-out$Haed
pd.out$Eaed <-out$Eaed
```

Now we have all the metrics calculated and can take a glimpse at how they are related to one another. Generally, in the literature, little is known about which metrics to use when, and which of them are redundant (but see a recently published review by Tucker et al. 2016). We can get a brief view into this by performing an ordination on these metrics using `metaMDS()` from the *vegan* package, and then plotting the results using `ordiplot()`. The normal data type for *vegan* would be a species by site matrix, but we can simply switch species to our calculated metrics:

```
library(vegan)

pd.mds <-metaMDS(pd.out,trace=FALSE)

ordiplot(pd.mds,type="t", display = "species")
```

There are some interesting patterns apparent in figure 3.11. Two regions in MDS space show multiple identical metrics (shaded squares in fig. 3.11). One of these, contains PSV and MPD as expected, as well as the intensive quadratic entropy, which is not surprising when we look at table 3.3; it is just another way of calculating a mean pairwise distance. These metrics do not give identical values; rather they are metrics based on the same input information—but scaled differently—that behave similarly. The other region that includes similarly behaving metrics includes PSE, MPD_{ab}, and the Rao's index. Remember PSE is the abundance-weighted version of PSV, so it makes sense that it is closely related to MPD_{ab}. The measures that are most unique in figure 3.11 include IAC, PAE, and the extensive quadratic entropy. The last measure is the sum of all pairwise measures in the distance matrix and is a difficult measure to interpret meaningfully. In this measure, internal edges are being summed multiple times, which may be statistically problematic.

So, what measures should we use? There is no single metric that should be selected above all others (see Tucker et al. 2016). One needs to balance multiple priorities, such as (1) ease of interpretation, (2) information content, (3) behavior across gradients, and (4) uniqueness relative to other measures (table 3.4). When deciding on metrics, these four characteristics

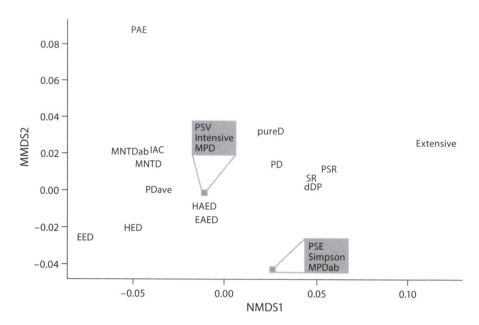

Figure 3.11. Similarity of metrics in multivariate space.

TABLE 3.4.
Considerations when selecting metrics to use.

Consideration	Example	Strategy
1) Ease of interpretation	Some measures are just easier to interpret. PD makes intuitive sense whereas extensive quadratic entropy is difficult to justify.	If some metrics appear mysterious, use those that are easier to understand or justify.
2) Information content	Some metrics do not contain the information the researcher needs to consider. If species abundance distributions are a critical part of the hypothesis test, then use abundance-weighted measures.	Understand what information is going into the metrics and whether this information is useful for hypothesis testing.
3) Behavior	Measures behave in very different ways. If we are looking at sites across a richness gradient, then one needs to carefully consider those that scale with richness to those that are independent of richness.	Use specific hypotheses or null models that inform the correct metric choice. Be aware of subtle issues like changes in variance across gradients.
4) Uniqueness	Some metrics relay similar information. There isn't much value in comparing PSV to MPD.	If comparing multiple metrics is desirable, select ones that are known to provide different or complementary information.

need to be considered, and choice of metric might vary with the question being asked of the data. However, there is nothing nefarious about examining multiple metrics and then using *post hoc* tests to explore differences between them. For example, if we are examining diversity patterns across an elevation gradient, but do not have a good *a priori* justification for which metrics should be most sensitive to this, it is an acceptable strategy to use a data mining approach to see how different metrics behave. Many good ecological papers uncover plausible hypotheses using inductive analyses on observational data. However, when using such an approach care must be taken not to simply pick and choose among alternative metrics until desired significance emerges. An alternative approach would be to select and compare metrics based on *a priori* groups (e.g., table 3.3 or Pearse et al. 2014) or mathematical families (e.g., Pavoine and Bonsall 2011), though we need to be confident that metrics exhibit similar behaviors within such groupings. More work is required to develop a framework to guide metric choice, such as the recent work by Tucker et al. (2016).

3.3. A NOTE ABOUT PHYLODIVERSITY MEASURES—MOVING FROM THE CAUSES TO THE CONSEQUENCES OF DIVERSITY

In this chapter we have gone over a substantial number of phylodiversity metrics, but we haven't really given a full account of what we might use them for. In the first half of this chapter, we explicitly considered measuring phylogeny to look for nonrandom patterns and link them with potential mechanisms of community assembly. However, it is not clear how or whether many of the metrics listed in table 3.3 can actually help us to understand community assembly directly. Many of these diversity measures can be used in comparison to traditional taxonomic diversity measures. For example, H_d and H_{aed} can be used in comparison to the Shannon index, and both the Hill numbers ($^qD(P)$) and Rao's entropy can be directly compared to their taxonomic equivalents.

Given that these phylodiversity measures quantify community diversity, we can articulate hypotheses and analyze them similarly to taxonomic diversity. This means that these measures can either be a dependent variable that is to be explained by some other variable, or an independent variable used to explain some other variable. So far, the vast majority of studies have focused on phylodiversity as the dependent variable—as something needing to be explained. For example, studies have examined how phylodiversity is influenced by natural gradients in environmental factors such as elevation, temperature, or precipitation (Bryant et al. 2008, Graham et al. 2009, Pellissier et al. 2013).

Less often tested is the how phylodiversity, as the independent variable, informs our understanding of other patterns. Within community ecology, there are many hypotheses testing ecological processes that are dependent on the fill of niche space. Therefore, any diversity measure can be used as the independent variable so long as it is correlated with the degree of niche filling. As an example, the invasions literature has a long tradition of asking whether species-rich communities are less easily invaded than those with fewer species because more diverse assemblages are thought to have fewer open niche opportunities for invaders (Elton 1958, Case 1990, Hector et al. 2001, Zavaleta and Hulvey 2007, Alexander et al. 2011). This reasoning has been extended to phylodiversity, where it is assumed that greater phylodiversity means that more of the niche space has been filled (Duncan and Williams 2002, Diez et al. 2008, Jiang et al. 2010, Thuiller et al. 2010). Of course testing the relationship between phylodiversity and invasion depends on the mechanisms at play and the relative importance of (1) niche differences that minimize competition, or (2) preadaptation represented by phylogenetically conserved niches (Daehler 2001, Proches et al. 2008, Thuiller et al. 2010, Li et al. 2015).

Another example of diversity as the independent variable is seen in research that links biodiversity to ecosystem function. Classic work on this linkage has explicitly assumed that a greater number of species means that on average more of the available niche space is occupied by species, and so more of the available resources can be utilized by species (Tilman et al. 1996, Tilman 1999, Naeem 2002, Balvanera et al. 2006, Cardinale et al. 2006, Zavaleta et al. 2010). Definitions of biodiversity in this work have been extended to include phylodiversity measures, and supply evidence that phylogenetic measures explain variation in ecosystem function more effectively than traditional taxonomic diversity measures (Maherali and Klironomos 2007, Cadotte et al. 2008, Cadotte et al. 2012, Cadotte 2013, Liu et al. 2015). The likely reason why phylodiversity appears strongly correlated with ecosystem function is because phylogenetic measures include niche divergence, with some species being more unique (and thus contributing more to ecosystem function), whereas taxonomic measures treat all species as equally different from one another (Mouquet et al. 2012, Srivastava et al. 2012).

Thus, phylodiversity measures can be used as either dependent or independent variables in analyses, but we should always be cognizant of the assumptions and ramifications of the different uses of phylodiversity. Further, and perhaps underappreciated, is the reality that multiple drivers affect patterns of diversity and, in turn, diversity influences a multitude of processes. While it therefore seems simple to say that we can use phylogenetic diversity as the dependent or independent variable, we need to ensure that we understand the complexity of diversity and realize that our inferences are limited by our assumptions.

3.4. CONCLUSION

There are a plethora of phylodiversity measures, and even more ways to use them. This is both a boon and a bane to ecology and conservation (Rosauer and Mooers 2013, Winter et al. 2013b, Tucker et al. 2016). We have reviewed many of the common types of community level phylodiversity measures and outlined how they handle phylogenetic information. But what remains unclear is which measure ought to be used for which question. We have tried to provide some guidance, but we require further development of frameworks linking measures of phylogenetic diversity to interpretations about the evolutionary and ecological processes from which they are derived.

While one could certainly calculate all available phylodiversity measures and assess statistical significance in analyses and then select the one that seems to be the most powerful, this approach lacks a real *a priori* sense of why particular measures are important. For example, some measures are more sensitive to terminal edges, while others capture more information from deeper edges, and we should be aware that the selection of a particular measure could subtly influence our interpretations of observed patterns.

~~~~~~

CHAPTER 4

Randomizations, Null Distributions, and Hypothesis Testing

Our ignorance of history causes us to slander our own times.
—*Gustave Flaubert*

How does one test if an observed pattern is significant when there are no standard statistical tests or known probability distributions associated with the data type? The answer is surprisingly easy—you build distributions from the data itself. Measures from phylogenetic data violate many of the basic assumptions of parametric statistics, such as independent errors, normal distribution, and homoscedastic or equal variance (e.g., Underwood 1997)—this last point, changes in variance, we will revisit in chapter 10. Phylogenetic data thus present serious challenges to traditional, parametric analyses. The first challenge is that phylogenetic data are nonindependent. Pairwise distances contain information from two species, and those species appear in some number of other pairwise measures, meaning those distances are not independent from one another. Secondly, due to the structure of phylogenetic topologies (i.e., a single deep split with more splits toward the tips of the tree) the distribution of values often follows very peculiar biased, multimodal, or truncated distributions. For example, from the Jasper Ridge data used in chapter 3, we can see that the distribution of pairwise distances is highly skewed toward large values because the node that contains the greatest number of pairwise distances is the basal split (fig. 4.1). Hierarchical objects will always have this type of distribution of pairwise distances. Finally, pairwise measures are not homoscedastic, meaning that the variance changes across a linear relationship. Pairwise measures are well known to exhibit decreasing variance across a gradient of species richness (fig. 3.8).

Data that violate the basic assumptions of parametric tests require a different statistical approach. Often researchers choose to employ distribution-free or nonparametric tests, which are free from assumptions based on *a priori* distributions (Edgington 1987, Manly 1991). With increased processor speeds, randomization or permutation tests, based on a null model,[1] have become commonplace (Edgington 1987, Manly 1991) and the preferred method of analysis for some areas of research (Pillar and Orloci 1996). Gotelli and Graves (1996, 3–4) define this type of null model approach as

> a pattern-generating model that is based on randomization of ecological data or random sampling from a known or imagined distribution. The null model is designed with respect to some ecological or evolutionary processes of interest. Certain elements of the data are held constant, and others are allowed to vary stochastically to create new assemblage patterns. The randomization is designed to produce a pattern that would be expected in the absence of a particular ecological mechanism.

1. The use of the term "null model" is commonplace in ecology. We will use "randomization test" since we wish to be specific about the procedure producing our null distribution.

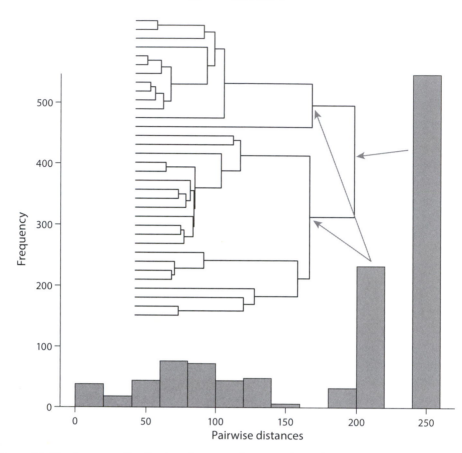

Figure 4.1. The frequency distribution of pairwise distances shows that they are often heavily biased toward large distances.

Randomization tests are conceptually quite easy to understand. They involve computing some value or test statistic from the data (e.g., mean distance or regression coefficient), and then calculating a null distribution or expectation of these values by randomizing the order of the data hundreds or thousands of times. Then we see where the observed value falls relative to the null distribution. If the observed value lies outside the 95% confidence interval,[2] then we conclude that the null hypothesis is not supported, and that our alternative hypothesis may be inferred. This would be analogous to a parametric test with $\alpha = 0.05$.

Here we will examine a nonphylogenetic example to establish the logic of a randomization test. Let's create two variables, a dependent variable and a predictor, which we know to be linearly related:

```
dat <-data.frame(x = rnorm(20,15))
dat$y <-dat$x + rnorm(20,0.5,0.5)
```

---

2. From here on we will use "95% CI" for the 95% confidence interval from the null distribution.

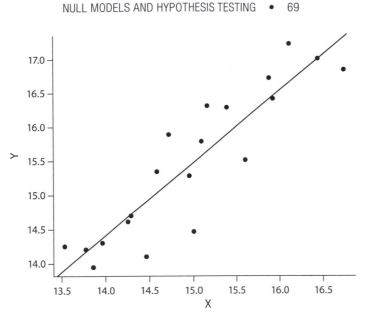

Figure 4.2. The relationship between two correlated variables (see text). The `abline()` function places a least-squares line on the plot based on the linear model (`lm()`) function. We can run a standard parametric linear regression, which confirms our expectation of a significant relationship between *x* and *y*.

This code simply produces two variables, x and y, where x is 20 values pulled from a normal distribution with a mean of 15, and y is x + a random number with mean of 0.5 and standard deviation of 0.5. If we plot these two variables, we can see they are related with some variation provided by the `rnorm()` function (fig. 4.2). Here is the code to generate figure 4.2:

```
plot(dat$x,dat$y,xlab="X",ylab="Y",pch=16)
abline(lm(dat$y~dat$x))
```

Further, a linear model confirms this strong relationship:

```
mod<-lm(dat$y~dat$x)
summary(mod)

Call:
lm(formula = dat$y ~ dat$x)

Residuals:
    Min       1Q    Median       3Q      Max
-1.00949 -0.18322 -0.01087  0.31966  0.71163

Coefficients:
            Estimate Std. Error  t value  Pr(>|t|)
(Intercept)  -0.4853     1.8018   -0.269      0.79
dat$x         1.0639     0.1200    8.867  5.49e-08 ***
---
Signif. codes: 0 '***' 0.001 '**' 0.01 '*' 0.05 '.' 0.1 ' ' 1
```

```
Residual standard error: 0.4826 on 18 degrees of freedom
Multiple R-squared: 0.8137,        Adjusted R-squared: 0.8034
F-statistic: 78.63 on 1 and 18 DF, p-value: 5.492e-08
```

What we need to focus on is the coefficient estimate (1.039), which has very low standard error (0.1200). This is the slope of the relationship between x and y. It is this coefficient we will use for comparison to a null model. We can pull this out of the *mod* object we created:

```
coef.obs<-mod$coefficients[2]
```

Now, if we wish to perform a randomization test (say we doubted the validity of some key parametric assumption) we will want to calculate coefficients from random data. We can use the `sample()` function, which takes a random sample from a vector. We will randomly sample *x*, and we will sample the same number of observations that *x* contains using `length(x)`. The default for `sample()` is to sample without replacement, meaning once a value is selected it is removed from the pool. We could have changed this by specifying `replace = TRUE`; however, we will not do this because if we allow the selection of values more than once, then we change the relative frequency of values and are thus altering two aspects of the data simultaneously (the correlation between *x* and *y*, and the frequency distribution of *x* values), making it difficult to develop inferences if our observed values fall outside the null expectation. Here is a simple function to return a linear regression coefficient from randomized values:

```
lm.rand <-function(x,y){
    r.coef <-lm(y~sample(x,length(x)))$coefficients[2]
    r.coef
      }
lm.rand(dat$x,dat$y)
```

This returns the coefficient from a linear model with a randomized x variable. For a real randomization test, we will need to perform this function multiple times (we will use 999). Here we will create a new function that uses the `replicate()` function to repeat a function and then calculate the *P*-value as the rank of the observed coefficient against the full distribution of random coefficients. This is how `ses.mpd()` and `ses.mntd()` from the *picante* and *pez* packages calculate their *P*-values (see chapter 3). In this function we will need to supply the x and y variables, along with the number of randomizations we want:

```
coef.rand.test <-function(x,y,N=999){
    coef.obs <-lm(y~x)$coefficients[2]
    coef.null <-replicate(N,lm.rand(x,y))
    coef.obs.rank <-rank(c(coef.obs,coef.null))[1]
    coef.obs.p <-coef.obs.rank/(N+1)
    if (coef.obs.p > 0.5) coef.obs.p <-(1-coef.obs.p)*2
    else coef.obs.p <-coef.obs.p*2
    data.frame(coef.obs,mean.coef.null = mean(coef.null),
        sd.coef.null = sd(coef.null),
        coef.obs.rank,coef.obs.p)
      }

coef.rand.test(dat$x,dat$y)
```

This returns the following (remember the values will differ every time because of the calls to `rnorm()` and `sample()`):

```
        coef.obs mean.coef.null sd.coef.null coef.obs.rank coef.obs.p
x       1.063947    0.01165480    0.258946          1000          0
```

From this output we can see that the observed coefficient is much larger than the null, and the standard deviation is fairly low. The resulting P-value is 0 (note that in the function we have an `if` statement which reports the P-values from a two sided test), and is thus outside the bounds of the 95% CI. Recall that the P-value from the parametric analysis above was $P = 5.49 \times 10^{-8}$; our finding that $P = 0$ for 999 randomizations is consistent with this result. We would need to run approximately 10,000,000 randomizations to get significant digits greater than 0 at 10^{-8} digits.

This simple example illustrates many of the basic statistical and programming tenets of randomization tests: (1) calculate the metric of interest on the observed data, (2) randomly sample without replacement—you don't want to sample the same number twice, (3) calculate the same metric on the randomized data, (4) repeat the randomization N times, and (5) test if the observed value is within the 95% CI of the null values.

As an aside, would we ever want to randomly sample with replacement? The answer is yes, and here we provide a couple of examples. If we are randomly assigning spatial locations to species, and more than one species can occur in a site, then we would want to sample sites multiple times. Similarly, if we were assigning individuals to species to construct some null abundance distributions, then again we might want to sample multiple times.

One issue to note is that randomization tests do have an important assumption; that is, if there are multiple treatments or groupings being randomized, they should come from the same population (Lunneborg 2000).[3] For example, if we were examining coexistence patterns, we wouldn't want to randomize community data from Borneo and Minnesota in the same analysis. The other issue is determining how many randomizations to perform. For time-consuming calculations, there may be computational limitations or time constraints, and so we would not want to run more randomizations than necessary and would need to make a decision about the number of randomizations to use, based on the total number of possible combinations. For two groups being pooled and randomized, the number of possible combinations is:

$$\frac{(n_1 + n_2)!}{n_1! \cdot n_2!} \tag{4.1}$$

and, for example, if we have two groups with 12 observations each, then there would be 2,704,156 possible combinations for two random groups with 12 observations each. What proportion of this number is a sufficient for a randomization test? There's no clear answer, but it partially depends on the variability in the data. A distribution with a low standard deviation will only require a small proportion of possible combinations in order to produce a good estimate of the mean and 95% CI. Most randomizations of relatively simple data often select 999 randomizations,[4] but more should be used to produce more accurate estimates.

3. Clifford Lunneborg published this article on his website at the University of Washington. He since passed away and the website was removed. Please contact the authors if a copy is desired.

4. Why do randomization tests seem to use 1 less than the logical number (e.g., 999 instead of 1,000)? The convention is more for elegance and doesn't affect interpretation—if you use 1,000 the results will be the same. The reason is that we often want to calculate our P as x/1,000, and since our observed value is an observation, the calculation of P is 999 + 1 (for our observed value) = 1000.

The number 4,999 would be good for most of our more complex data sets.[5] The tradeoff of course is that phylogenetic analyses are often time consuming and randomizations can take an enormous amount of computational time. For example, randomizations for some phylogenetic signal tests in chapter 5 can take days to complete; these involve 5,000 permutations for phylogenetic trees with hundreds of species. And these large trees are the type of data that require many permutations.

Null models in ecology have grown from simple beginnings to a widely used approach for testing for nonrandom ecological patterns. Before discussing null modeling for phylogenetic tests it is worth understanding some of this history.

4.1. A BRIEF HISTORY OF RANDOMIZATION TESTS IN ECOLOGY (OR THE SIMBERLOFFIAN SHIFT IN ECOLOGY)

Some of the most important scientific controversies in ecology have involved, in one way or another, randomization tests (e.g., Jarvinen 1982, Gotelli and Graves 1996). Here we discuss two important academic conflicts that highlight the need for using randomizations in ecological analyses.

4.1.1. Species-Genus Ratios and Competition between Close Relatives

One of the earliest controversies in ecology revolved around the numbers of species per genus on islands and inferences of competition. As we noted previously, Darwin (1859) originally suggested the following:

> As species of the same genus have usually, though by no means invariably, some similarity in habits and constitution, and always in structure, the struggle will generally be more severe between species of the same genus, when they come into competition with each other, than between species of distinct genera.

To test this hypotheses, the Swiss botanist, Paul Jaccard (1901) created a "generic coefficient" to describe biogeographical patterns and measure the effects of competition on diversity. The generic coefficient was a form of the species-genus ratio (S/G), calculated as $G/S \times 100$, and he interpreted a low S/G ratio (or high coefficient) to mean that competition between close relatives was high, and a high ratio (low coefficient) meant that there was a high diversity of "ecological conditions" supporting closely related species in slightly different habitats (Jaccard 1922). At the same time as Jaccard was working on his generic coefficient, the Finnish botanist Alvar Palmgren compiled S/G patterns across the Aland Islands and inferred the low S/G values on distant islands to reflect random chance caused by stochastic assembly (Palmgren 1921). Over several years, Jaccard and Palmgren had a heated exchange in the literature about interpreting S/G ratios (e.g., Jaccard 1922, Palmgren 1925). Palmgren's contention was that the S/G ratios he observed were related to the number of species occurring on the islands—an argument that later work vindicates. A few years after their exchange, another Swiss scientist, Arthur Maillefer, showed that Jaccard's interpretation was not supported by statistical inference (Maillefer 1928, 1929). Maillefer

---

5. For many of our examples we will use 999 randomizations to keep computational time to a minimum while still providing relatively powerful tests.

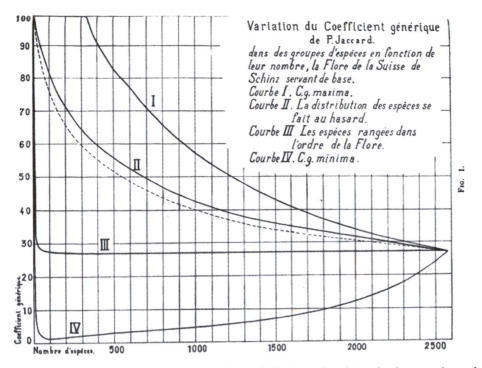

Figure 4.3. Jaccard's generic coefficients plotted by Maillefer shows the relationship between the coefficients (calculated as genera/species × 100) and species richness (Maillefer 1929). The four curves depict different scenarios. *Curve I* shows the maximum values possible, and *curve IV* is the minimum. *Curve III* occurs when coefficients are calculated with sampled values from a flora, which stays on a mean value. *Curve II* represents perhaps one of the first null models in ecology, where species are randomly sampled (*hasard* is translated as chance or luck) and the coefficient was calculated from the random assemblages.

created what is likely one of the first null models in ecology (Jarvinen 1982). He calculated the expected relationship between Jaccard's generic coefficient and species richness from "chance" communities that were randomly assembled (fig. 4.3, curve II). Maillefer rightly concluded that since the number of genera increases at a slower rate than richness, the ratio between the two couldn't be independent of richness.

This example is especially pertinent because it foreshadowed another debate 20 years later, and not just in terms of using a null expectation; the question was whether S/G ratios could be understood without comparison to the appropriate null. Elton (1946) examined an impressive set of studies to show that small assemblages tended to have low S/G ratios, which he thought indicated competitive interactions. Mirroring the earlier debate, Williams (1947) showed that S/G ratios were not independent of richness and that inferences about competition can only be supported if observed S/G values differed from expected null values. However, the error of inferring competition from S/G ratios without comparing them to null expectations continued into the 1960s (Grant 1966, Moreau 1966). Then, Dan Simberloff (1970) showed, unambiguously, that, independent of any ecological mechanism, lower S/G ratios are expected on islands with fewer species. Because he compared observational values to null expectations, Simberloff was able to show that assemblages actually tended to have higher S/G ratios than one would expect by chance (Simberloff 1970). So not only is competition not supported, but the available evidence indicated that perhaps there were more closely related species on islands, which Simberloff interpreted as close relatives preferring the same environments (Simberloff 1970).

4.1.2. Co-occurrence Patterns and Competitive Coexistence

The need for null models became very apparent by the 1970s. Donald Strong would go so far as to say many ecological inferences in the 1960s and 70s were incorrect because they did not explicitly consider a correct and meaningful null model (Strong 1980). Perhaps the best-known ecological controversy relating to null models and statistical inference occurred in the mid-1970s in a debate over species co-occurrence patterns. In his detailed natural history study of the birds of New Guinea and the surrounding islands, Jared Diamond (1975) concluded that competitive interactions underpinned species diversity and co-occurrence patterns. He showed that there were subsets of species that formed combinations that could not be invaded on small islands, and these were subsets of those on larger islands. Further, he determined that there were "forbidden" combinations of species that could not coexist with one another. Diamond's observations led him to formulate the following rules of community assembly (Diamond 1975, 423):

1. If one considers all the combinations that can be formed from a group of related species, only certain ones of these combinations can exist in nature.
2. Permissible combinations resist invaders that would transform them into forbidden combinations.
3. A combination that is stable on a large or species-rich island may be unstable on a small or species-poor island.
4. On a small or species-poor island, a combination may resist invaders that would be incorporated on a larger or more species-rich island.
5. Some pairs of species never coexist, either by themselves or as part of a larger combination.
6. Some pairs of species that form an unstable combination by themselves may form part of a stable larger combination.
7. Conversely, some combinations that are composed entirely of stable subcombinations are themselves unstable.

However, Diamond's assembly rules do not compare observed co-occurrence patterns to those expected from random sampling. This led Simberloff to opine: "I suggest that if competitive interactions did not exist most published biogeographic distributions would be similar to those now observed, which is tantamount to arguing that competition does not greatly influence island colonization" (Simberloff 1978, 715).

Connor and Simberloff (1979) took a more aggressive stance when pointing out that Diamond's observations do not naturally evoke a specific mechanism without further evidence. In discussing Diamond's assembly rules, Connor and Simberloff suggested that they could "show that every assembly rule is either tautological, trivial, or a pattern expected were species distributed at random" (1979, 1132). The reason to doubt Diamond's assertions is that "to demonstrate that competition is responsible for the joint distribution of species, one would have to falsify a null hypothesis stating that the distributions are generated by the species randomly and individually" (Connor and Simberloff 1979, 1132).

Connor and Simberloff, unable to obtain Diamond's original data, analyzed three other data sets; one was New Hebrides birds and the other two (birds and bats) were from the West Indies (Connor and Simberloff 1979). They examined the number of islands shared among pairs of species and compared these values to a random expectation (fig. 4.4). They used randomizations of the data to construct the null, subject to three constraints:

1. For each island, there is a fixed number of species; namely, that which is observed.
2. For each species, there is a fixed number of occurrences; namely, that which is observed.
3. Each species is placed only on islands with species numbers in the range for islands which that species is, in fact, observed to inhabit. That is, the "incidence" range convention is maintained, such that species that are observed only on very species poor islands are not placed on species rich islands (Connor and Simberloff 1979, 1133).

The first constraint maintains the richness of islands, and thus whatever limits the number of species (space, resources, etc.) is maintained. The second maintains species occupancy patterns—thus, rare species are kept rare and common species common. The final constraint may seem a little idiosyncratic, but Diamond's (1975) original analysis

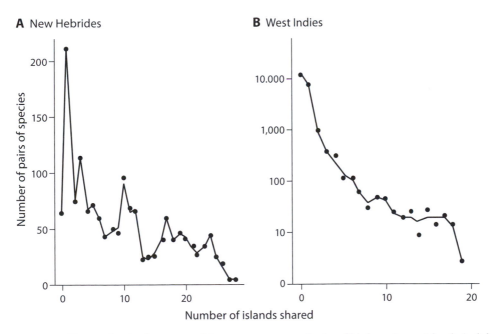

Figure. 4.4. The randomized patterns of the co-occurrence of pairs of bird species on islands (*solid lines*) compared to the observed number (*black dots*). Adapted from Connor and Simberloff 1979, p. 1134.

classified species into incidence groups (e.g., those that occur on species-rich islands only) to make inferences about the ecological mechanisms influencing their occupancy patterns. Connor and Simberloff showed that the observed number of islands on which pairs of species occur together was not different than that expected by their randomization (fig. 4.4; see also Connor and Simberloff 1986), though they did find statistically more nonoverlapping pairs of West Indies birds and West Indies bats then expected by chance (Gotelli and Graves 1996).

There were a number of criticisms of Connor and Simberloff's null model approach (e.g., Alatalo 1982, Gilpin and Diamond 1984), as well as further exchanges in the literature (e.g., Connor and Simberloff 1984); for a good summary, see Gotelli and Graves (1996). The principal concerns were threefold: the null models did not define species groups correctly, the constraints were overly restrictive, and the significance tests were incorrect (Gotelli and Graves 1996). These legitimate issues highlight the need to carefully consider how one should randomize community data.

4.2. HOW TO BUILD NULL COMMUNITIES

Community data can be characterized in a number of meaningful ways. If we have a list of communities, we may wish to know something about community structure. This might entail calculating the numbers of species in each community (richness), the number of communities each species is found in (occupancy), the number of individuals or total biomass or cover for each species (abundance), and the abundance distributions within communities. When we randomize community data, we should be cognizant of what aspects of this community structure we are affecting and what that means for inferring specific mechanisms (Harvey et al. 1983, Gotelli and Graves 1996). There are several sources that explain different approaches to randomizing community data (e.g., Gotelli and Graves 1996, Gotelli 2000, Miklós and Podani 2004), and we will not go over all the different randomization strategies here, but rather examine how to randomize a community matrix.

The basic question is: "What do you randomize and what do you fix?" We start with a community data matrix (table 4.1) where cells contain abundances, rows are communities

TABLE 4.1.
Sample site by species community matrix.

	Sp1	Sp2	Sp3	Sp4	Sp5	Sp6	Richness	Abundance
Com1	10	0	0	0	5	0	2	15
Com2	0	3	4	0	0	9	3	16
Com3	0	0	8	6	0	0	2	14
Com4	7	0	9	0	0	0	2	16
Com5	5	3	2	2	1	2	6	15
Occupancy	3	2	4	2	2	2		
Abundance	22	6	23	8	6	11		Total = 76

TABLE 4.2.
Gotelli's nine possible community matrix randomizations (Gotelli 2000).

| Constraint | Constraint | | |
	Columns equiprobable	Columns proportional	Column sums fixed
Rows equiprobable	All species and sites are equiprobable. All rearrangements equally likely.	All sites equiprobable. Probability of species occurrence proportional to observed occurrence frequencies.	Species occurrence totals fixed. All sites are equiprobable.
Rows proportional	All species are equiprobable. Probabilities of occurrence in sites proportional to observed richness.	Probabilities of species occurrence are conditional on both site and species totals.	Species occurrence totals fixed. Probabilities of occurrence in sites proportional to observed richness.
Row sums fixed	All species are equiprobable and species number per site fixed.	Site richness is fixed. Probability of species occurrence proportional to observed occurrence frequencies.	Cannot be simulated by filling an empty matrix— complex algorithms required.

or sites, and columns are species.[6] If we wish to retain the information that sites differ in the number of species they can hold (i.e., different sizes or resources) then we would want a randomization that maintains row totals. However, if one simply swaps cell values, while retaining row totals, species occupancies and community abundance will change. This may be fine for the particular hypotheses being tested, and enforcing too many constraints on the randomization will make rejecting it for an alternative hypothesis impossible. Gotelli (2000) provides the full universe of nine possible randomization procedures on community data matrices (see table 4.2). Further, we could decide to change the abundance distributions within communities by assigning individuals to species at random or following a known distribution (Gotelli and Graves 1996).

We will go through a simple example where we randomize a matrix. Here is the code to create the matrix that is shown in table 4.1:

```
com.dat<-matrix(c(10,0,0,0,5,0, 0,3,4,0,0,9, 0,0,8,6,0,0,
        7,0,9,0,0,0, 5,3,2,2,1,2),nrow=5, byrow=TRUE)
rownames(com.dat)<-c("Com1","Com2","Com3","Com4","Com5")
colnames(com.dat)<-c("Sp1","Sp2","Sp3","Sp4","Sp5","Sp6")
```

To calculate abundance, occupancy, and richness totals we will use functions from the family of `apply` functions; `apply` functions generally run a function across some set of data. They run faster and require less code than a "for" loop but can be difficult to decipher and less flexible. As a general rule, if there is a problem that requires us to run the same function across elements of a vector, columns of a matrix or data frame, elements of a list, or even multiple data sets, we should look at using an `apply` function. If we type `??apply`, we see that there are a multitude of functions that use the word "apply" somewhere in the description. We are interested in the ones in the base package. Table 4.3 shows the different `apply`

6. Please note that other sources (e.g., Gotelli 2000) have rows as species and columns as sites.

TABLE 4.3.
The apply functions in R, the input data objects, and what gets passed through the function.

Function	Input object	Pass
apply	Matrix, data frame, or array	Single rows or columns
lapply	List	List component
sapply	Vector	Single element
tapply	Vector	Vector, by groups
by	Data frame	Rows
aggregate	Vector and grouping variable	Groups

functions and details what information goes into them, and what gets passed through the function. The data that gets passed can be thought of as what the function is repeating across. For example, if we calculate means for the columns of a data frame, it is the columns that are being passed through the function.

In general, apply functions take this basic form (with the '_' to indicate that we could be using any of the apply functions in table 4.3):

_apply(dat,'other arguments', function)

The *dat* is the input object (e. g., a vector) and 'other arguments' is specific to the particular apply function we are using (e.g., we need to specify a column of the data frame that contains the grouping variable in tapply). Finally, we need to specify the function—that is, what we want to calculate. If we pass a prescribed function, such as mean, we simply type the function name without parentheses. If we wish to implement a more tailored calculation we may need to create a function, but we need only include x for the argument, since the apply function assumes that what is being passed to the function will be x. For our example we will calculate a couple of things, one of which is counts of the number of species. There are functions that do this in various packages, but we will count the number of cells in a row or column only if their value is greater than 0. Using x, which will appear in the apply function, we can count the number of species or occurrences as

length(x[x>0]]

This returns the length of elements in x that are greater than 0.

Since we wish to sum row and column totals from our community matrix (table 4.1), we will use the apply function, and we need to specify the MARGIN. This tells the apply function to go across rows (MARGIN = 1) or columns (MARGIN = 2):

apply(com.dat,MARGIN=1,function(x) length(x[x>0]))

This returns:

```
Com1  Com2  Com3  Com4  Com5
   2     3     2     2     6
```

This is the species richness for each community. To calculate species' occupancy we change the second argument to MARGIN[7] = 2:

---

7. Note that we do not actually have to type out MARGIN in the apply function as it is always the second argument in the function.

```
apply(com.dat,2,function(x) length(x[x>0]))
```

This returns:

```
Sp1  Sp2  Sp3  Sp4  Sp5  Sp6
  3    2    4    2    2    2
```

This is the number of communities occupied by the different species.

Now we can also calculate community abundance and total species abundances using the sum function.

Community abundance is entered as

```
apply(com.dat,1,sum)
```

Species abundances is entered as

```
apply(com.dat,2,sum)
```

Community 5 stands out in the number of species it contains, and perhaps we want to know if this number is significantly different than for the other communities. The simplest randomization would be to swap all the cell values without constraining the row or column totals (i.e., the first cell in table 4.2). We use the sample() function; if no other arguments are specified it will return a vector with all of our cell values randomized, and sampled without replacement. Since this function call returns a vector, we need to turn it back into a matrix to perform our calculations:

```
com.tmp<-matrix(sample(com.dat),nrow=5)
```

Each time we run this, the cells' values will be in different locations (which is what we want). To compare the observed maximum number of species (6) to the maximum for random assemblages we will use a "for" loop, applied here across 999 randomizations:

```
out<-NULL
for (i in 1:999){

out[i]<-max(apply(matrix(sample(com.dat),nrow=5),1,
       function(x) length(x[x>0])))

}
```

Here we are simply using the apply function on the randomized matrix, and using the max() function to return the maximal richness 999 times. Thus the object *out* is a vector with 999 values and we want to know if our observed maximum richness is significantly different than this. The use of null models in the 1970s often used a statistical test, like a Chi-square test (e.g., Connor and Simberloff 1979), but it can't be assumed that the randomized data conform to test statistic assumptions. However, there is a more direct way to assess significance (Veech 2012): we can compare the observed value directly to the randomized values (fig. 4.5), as we also describe in chapter 3.

To do this, we use the same approach that we used in the opening example of this chapter; that is, we use the rank() function. First we will store the maximum observed richness:

```
max.obs<-max(apply(com.dat,1,function(x) length(x[x>0])))
```

And now we find its rank relative to the null values and calculate the *P*-value:

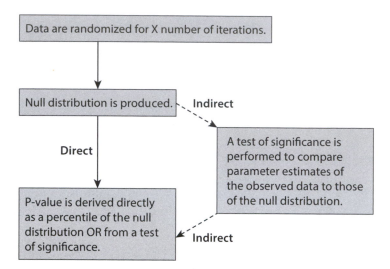

Figure 4.5. The anatomy of a randomization test showing that comparisons to the null distribution are more direct than performing a statistical analysis on the observed and null data. Adapted from Veech (2012).

```
obs.rank<-rank(c(max.obs,out))[1]
P_value<-obs.rank/(length(out)+1)

(1-P_value)*2
```

Since the distribution is two tailed, we subtract the *P*-value from 1 when it is larger than 0.5 in order to get a one-tailed value. In our case it is 0.033—again this value will change marginally each time the randomization is run. Thus, we would reject the null hypothesis (that the observed maximum richness is no different than random) for the alternative that community 5 has more species than we expect by chance.

Looking at the histogram of randomized values (fig. 4.6), we can see that a maximum richness of 6 is a rare occurrence. We can also compare the observed to the mean null value (μ) and 95% CI:

```
mean(out)
quantile(out,c(0.025,0.975))
```

This gives us $\mu = 4.41$, 95% CI = 4, 6; thus our observed value is just at the outer limit of the 95% CI. Given that we get different values, we can assess the robustness of observing a *P*-value of 0.033, given 999 randomizations, by rerunning this randomization itself 1,000 times to create a distribution of *P*-values like this:

```
out.p<-NULL
for (j in 1:1000){

out<-NULL

for (i in 1:999){

out[i]<-max(apply(matrix(sample(com.dat),nrow=5),1,
    function(x) length(x[x>0])))
```

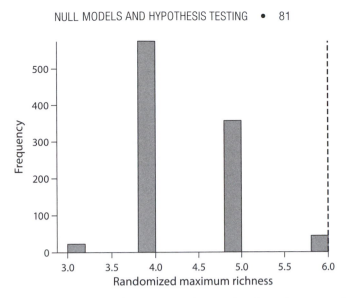

Figure 4.6. A histogram of randomized maximum richness values from the community data matrix shown in table 4.1. The dashed line indicates our observed maximum richness.

```
}
obs.rank<-rank(c(max.obs,out))[1]
coef.obs.p <-obs.rank/(length(out)+1)
if (coef.obs.p > 0.5) out.p[j] <-(1-coef.obs.p)*2
else out.p[j] <-coef.obs.p*2

}
hist(out.p,xlab="Randomized P-values",main=NULL)
abline(v=0.05,lty="dashed",col="orange",lwd=2)
```

The histogram of *P*-values is shown in figure 4.7, and there is a substantial tail larger than 0.05. This would indicate that the 999 randomizations in our original test weren't high enough; in this case we may want to go to 9,999 randomizations.

To implement different constraints on the randomization procedure would require writing more complex functions. Fortunately, *picante* has a handy function `randomizeMatrix` that implements four different constrained randomizations. The first, and the default randomization, maintains species frequencies, where the column sums remain constant, and just the vertical location of the abundance values change (i.e. species occurrence frequencies are maintained). Here is an example:

```
randomizeMatrix(com.dat,null.model="frequency")
```

If we run this a few times, we would see that in each matrix, the cell values are simply moving vertically. Later on we will use this randomization to construct a null distribution. The next constrained randomization maintains sample richness but not occurrence frequency:

```
randomizeMatrix(com.dat,null.model="richness")
```

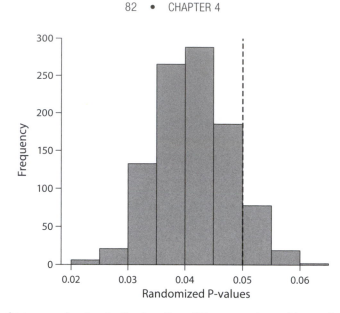

Figure 4.7. The histogram showing the *P*-values from 999 permutations of the randomization test in figure 4.6; the *dashed line* indicates *P* = 0.05. Given the relatively wide variance in *P*-values, we may want to rerun the previous randomization test with a greater number of randomizations.

The next two are fully constrained, keeping both richness and occurrence frequency constant. The first, `independentswap`, is based on the algorithm by Gotelli (2000), which finds 2 × 2 submatrices that can be swapped and retain the constraints. The second, `trialswap`, is similar to `independentswap` but is able to produce an unbiased series of randomized matrices (Miklós and Podani 2004). Here is the implementation in *picante*:

```
randomizeMatrix(com.dat,null.model="independentswap")

randomizeMatrix(com.dat,null.model="trialswap")
```

We don't really care what a single randomization looks like, but rather, we wish to compare our observed maximum richness to a null *distribution*. It would obviously be trivial to do this with a randomization procedure that maintains row sums (richness), because we are preserving the observed value. We will look at randomizations assuming the first constraint, maintaining species frequencies of occurrence:

```
out.f<-NULL

for (i in 1:999){

out.f[i]<-max(apply(randomizeMatrix(com.dat,null.model=
      "frequency"),1,function(x) length(x[x>0]))))

}

obs.rank.f<-rank(c(max.obs,out.f))[1]
P_value.f<-obs.rank.f/(length(out.f)+1)

(1-P_value.f)*2
```

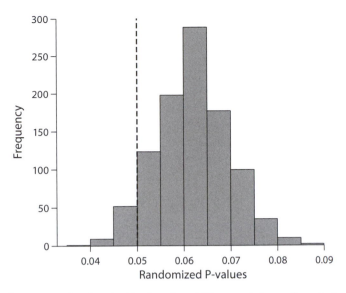

Figure 4.8. The histogram showing the *P*-values from 999 permutations of constrained randomizations that maintain species' frequencies of occurrence. The *dashed line* indicates *P* = 0.05. This case differs from figure 4.7 in that the null hypothesis would not be rejected.

Here we now return *P* = 0.064 (again, this will vary each time we run it). And if we look at the distribution of *P*-values (code not shown) (fig. 4.8), we can see that we lack evidence to reject the null hypothesis. This is an important point; by including constraints we change the power to reject a null hypothesis. We will be revisiting this.

4.3. RANDOMIZING PHYLOGENETIC DATA

The search for phylogenetic community patterns has featured prominently in the ecological literature over the past decade. This work has largely focused on detecting patterns of over- or underdispersion in local communities relative to a species pool (Webb 2000, Webb et al. 2002, Cavender-Bares et al. 2009, Mouquet et al. 2012). The reason Cam Webb's (2000) paper on the phylogenetic structure of Borneo tree communities had such a large impact was, in part, because he defined a metric of relatedness that used a null distribution to assess significance. Many phylogenetic analyses require randomization tests for hypothesis testing. Various functions implement a default randomization procedure in a specific function (e.g., `ses.mpd()` in *picante*, but *pez* also performs this analysis); however, it is important to consider what the implications of this choice are for detecting specific mechanisms (Hardy 2008, Mouquet et al. 2012).

What we need to randomize depends on the data and the hypotheses being tested. In chapter 3 we ran simple randomizations on a couple of measures without too much thought. We did not specify any constraints in the randomization tests we ran in chapter 3 (i.e., `ses.pd`, `ses.mpd`, `ses.mntd`), and the functions all default to a tip-swapping randomization, where the tip labels on the phylogeny are randomized and then matched to the community or trait data sets (fig. 4.9). (The different randomization routines will be

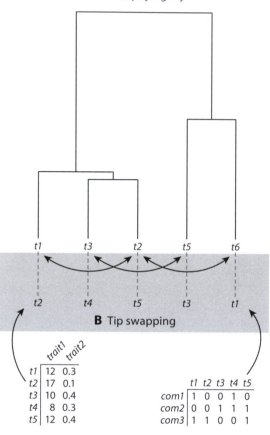

Figure 4.9. How tip-swapping randomizations work. For each iteration of the randomization we start with the real phylogeny (*A*), then randomize taxon names or tip labels (*B*), and then match the new randomized tip labels to the taxon names from the unchanged community or trait data. We would then calculate whatever metric or test statistic we were interested in, repeat some number of times (e.g., 999), and compare the distribution of values to the test statistic or metric from the empirical data.

discussed later.) Tip swapping is a randomization that simply switches species identities, allowing the frequency distribution of abundances and occupancies to remain unchanged; inferences from this procedure need to be carefully considered to ensure biological mechanisms can be reasonably inferred (e.g., De Bello 2012).

Here we will revisit these tests from chapter 3 and examine the randomization in greater detail. Recall that we ran the function `ses.mpd()` on the example *phylocom* data in *picante*. The function `ses.mpd()` calculates the standard effect size of the mean pairwise distance, relative to the mean and standard deviation from a series of null values:

```
data(phylocom)
out.mpd<-ses.mpd(phylocom$sample,cophenetic(phylocom$
    phylo),runs=999)
```

Let's focus on one particular assemblage called "Clump2b" (see fig. 3.5). We can pull out our metrics for this particular assemblage, which is the third community down in the matrix:

```
out.mpd[3,]
```

This returns (we have split the output for readability):

	ntaxa	mpd.obs	mpd.rand.mean	mpd.rand.sd
clump2b	8	7.142857	8.323109	0.3279291
	mpd.obs.rank	mpd.obs.z	mpd.obs.p	runs
	5	-3.599106	0.005	999

This shows that the assemblage is significantly clustered (z-value = −3.599, P-value = 0.005). Since this result was produced by the tip-swapping method, we can ask how robust the result is to the randomization method by performing the significance test on the four other methods discussed in the previous section:

```
out.mpd.f<-ses.mpd(phylocom$sample,cophenetic
        (phylocom$phylo),runs=999,null.model="frequency")
out.mpd.r<-ses.mpd(phylocom$sample,cophenetic
        (phylocom$phylo),runs=999,null.model="richness")
out.mpd.i<-ses.mpd(phylocom$sample,cophenetic
        (phylocom$phylo),runs=999,null.model="independentswap")
out.mpd.t<-ses.mpd(phylocom$sample,cophenetic
        (phylocom$phylo),runs=999,null.model="trialswap")
```

We can visualize the output by creating a data frame containing the results for the assemblage of interest, and pulling out z-values and P-values matching to Clump2b:

```
to.plot<-data.frame(Constraint=c("Tip-swap","Frequency",
        "Richness","Independent swap","Trial swap"),
    Z_value=c(out.mpd[3,6],out.mpd.f[3,6],out.mpd.r[3,6],out.
        mpd.i[3,6],out.mpd.t[3,6]),P_Value=c(out.
        mpd[3,7],out.mpd.f[3,7],out.mpd.r[3,7],out.
        mpd.i[3,7],out.mpd.t[3,7]))
```

This gives us the following data frame:

	Constraint	Z_value	P_Value
1	Tip-swap	-3.599106	0.0050
2	Frequency	-1.377138	0.0830
3	Richness	-3.303815	0.0050
4	Independent swap	-1.670930	0.0615
5	Trial swap	-1.712087	0.0590

This table shows differences in both z-values and P-values. If we graph these values, it is helpful to order the constraints, because R will plot them in alphabetical order. We can view their order by:

```
print(levels(to.plot$Constraint))
```

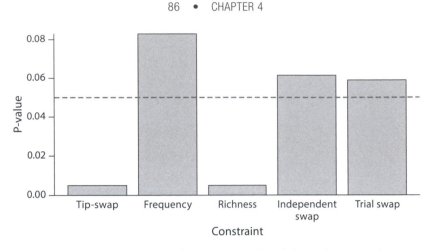

Figure 4.10. The *P*-values for the "clump2b" assemblage in the *Phylocom* data using the `ses.mpd()` function for different randomization procedures. The *dashed line* indicates *P* = 0.05. Three of the tests fail to reject the null hypothesis at *P* = 0.05.

This returns the following order of factors:

```
[1] "Frequency"    "Independent swap"    "Richness"
    "Tip-swap"     "Trial swap"
```

We can set the order with the `tip-swap` first and then in the order that we ran the tests:

```
to.plot$Constraint<-factor(to.plot$Constraint,levels
    (to.plot$Constraint)[c(3,1,4,2,5)])
```

Plotting these *P*-values reveals fairly substantial variation (fig. 4.10), but more importantly, some of the randomization methods result in nonsignificance for an assemblage that was generated so as to be underdispersed. There should be an informative way to proceed, other than haphazardly selecting a randomization procedure that produced a significant result. We could select a method that we think best informs our hypothesis. But what if we lack an *a priori* reason to select a method? It could be that these different methods encapsulate the extreme *P*-values, and so we could use some sort of model averaging. The simplest averaging method is to simply take the mean of the *P*-values. From our analysis, we would conclude that this assemblage is significantly underdispersed (average *P*-value = 0.0427), but it is much more preferable to select a specific null model that best matches our hypothesis or inference.

Here is the code to plot figure 4.10:

```
barplot(height=to.plot$P_Value,names=to.plot$Constraint,
    xlab="Constraint",ylab="P-value")
abline(h=0.05,lty="dashed",lwd=2,col="grey65")
```

4.3.1. Randomization Tests to Control for Richness Effects

Phylogenetic randomization can also be used to account for the fact that many phylogenetic measures are not independent of species richness (e.g., fig. 3.8). Just as in the species-genus ratio debates, it is difficult to make inferences about phylogenetic values without controlling

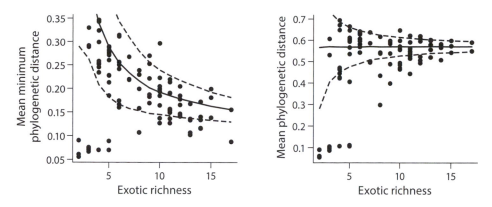

Figure 4.11. The relationship between richness and MNND or MPD. The *solid lines* indicate mean values from 999 randomizations and the *dashed lines* give the upper and lower bounds of the 95% CI. Adapted from Cadotte, Borer, et al. (2010).

for richness if they also co-vary with richness. In figure 3.8, both the mean and variance of phylogenetic measures varies across richness values. We often think about controlling for the mean effect (e.g., by analyzing the residuals of a PD-richness regression), but we seldom control for nonindependence in variance; in this case, using residuals may not solve the problem. For example, Cadotte, Borer, et al. (2010) found that exotic assemblages in California with very few species were much more likely to have MPD and MNTD values below the 95% CI from null models, indicating significant clustering (fig. 4.11).

This nonindependence becomes especially important if we wish to examine phylogenetic diversity patterns across some environmental gradient that may also influence species richness (e.g., fig. 4.12). Without correcting for richness, any relationship between phylogenetic diversity and elevation may be spurious and result from the fact that high elevation sites host fewer species. High elevation sites may be smaller in area or have fewer resources, which may simply contain smaller, but random, subsets of lower elevation communities (Alexander et al. 2011). Whereas, if we naively interpret patterns in figure 4.12A, we may conclude that higher elevation assemblages contain more closely related species, perhaps because of environmental filtering.

However, with a randomization test, we may see meaningful deviations from the null expectation. In figure 4.12C, the three sites with the fewest species (i.e., high elevation sites) fall below the 95% CI of the randomization (dashed lines). Thus, while the overall relationship between PD and elevation can be largely explained by the decline in richness across the gradient, there do appear to be communities at the highest elevation that contain species more closely related than expected by chance.

In an informative study, Helmus et al. (2010) examined the effect of disturbance on zooplankton communities in the northern United States and southern Canada (fig. 4.13). Disturbance reduced both the richness and phylogenetic diversity of the lake communities, and even though their measure of phylogenetic diversity (PSV) is mathematically independent of richness, one may still wonder if there is a relationship between the two (remember that PSV and MPD are statistically equivalent). The results of their analyses show that the decline in richness and phylogenetic diversity appear unrelated to one another. The community that lost the greatest number of species showed little change in

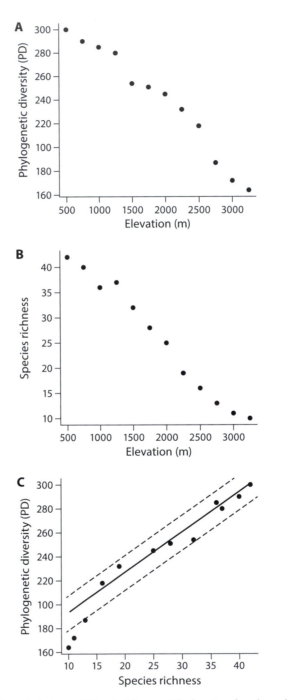

Figure 4.12. The relationship between PD and richness with elevation for a hypothetical data set. The correlation between richness and phylogenetic diversity may undermine PD-elevation relationship.

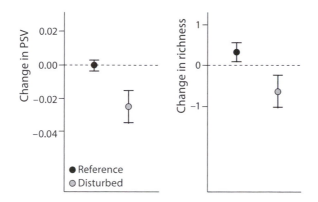

Figure 4.13. The effect of disturbance on lake zooplankton communities. *Gray circles* indicate disturbed lakes and *black circles* are from paired reference sites. While disturbance appears to reduce richness and the phylogenetic measure (on PSV see chapter 3), they are actually uncorrelated. Adapted from Helmus et al. (2010).

PSV, and some of the communities with the greatest PSV declines showed little change in richness (fig. 4.13).

4.3.2. Randomizations to Test Trait Data

So far, we have dealt with community data in terms of presence-absence and abundances, but ecolophylogenetic analyses often also involve species' traits. We often want to know whether communities are nonrandom in terms of traits. And if there is phylogenetic signal in the key traits, does it explain phylogenetic community patterns? While we do not explicitly explore trait-based measures of functional diversity in this book (please see Swenson 2014), we examine the phylogenetic signal in traits that are ecologically important and we use patterns of trait evolution to explain phylogenetically nonrandom community patterns. Chapter 5 explores how we can test for phylogenetic signal in traits and provides a brief overview of trait evolution.

Randomizing trait data is much more straightforward than randomizing community matrices. That said, we could potentially create null trait distributions from hypothesized models of, say, niche overlap. In that case, we could come up with increasingly complex models. For our purposes here, we will assume that we only want to randomize our observed trait values. The most frequently used approach is tip swapping (e.g., fig. 4.9). The basic idea again is that the tip labels are randomized and then matched to the trait data. In this case, species x will always have the same trait value but its location in the phylogeny will be randomized. This makes sense for phylogenetic signal tests, since we wish to compare either the phylogenetic variance-covariances of traits, or the reconstructed internal nodes to those expected by randomly distributing the traits (see the following chapter for further details).

For completeness, we can also randomize the trait matrix itself. We will use the trait data included in the *phylocom* data in *picante*. There are a number of ways to randomize this trait data. The easiest is to just randomize the rownames, while not changing the data. We can do this as follows:

```
tmp.traits<-phylocom$traits
rownames(tmp.traits)<-sample(rownames(phylocom$traits))
tmp.traits
```

Now, when we match the tip labels from the phylogeny to this data set in order to calculate the phylogenetic signal, the taxa are represented by different traits. This approach is equivalent to tip swapping. Chapter 5 describes how we might set up a significance test of phylogenetic signal using a similar approach. Be aware that we have retained the correlations among the different traits, and if we had reason to uncouple these, we would need to randomize the individual traits as vectors.

4.3.3. Altering the Tree

In the preceding sections we have randomized community data with a number of constraints, but we did not alter the phylogenetic tree itself. When we randomize the community matrix we are, in essence, asking: How likely are co-occurrence patterns given a tree? And to do this we randomize the community data. It can be just as valid to ask: Are co-occurrence patterns sensitive to speciation history? To assess this question we need to alter the tree.

The phylogenetic tree itself can be altered in a number of ways, and we might consider swapping the tip labels as one approach for which topological constraints are enforced. Swapping the tip labels assumes that the phylogenetic topology or the variance-covariance matrix remains unchanged. However, there are three theoretical ways to alter the tree structure itself (table 4.4). How we alter the phylogeny will influence our inference. For example, Cadotte (2015) uses several different phylogenetic randomizations to conclude that the ecological pattern he was studying (the relationship between phylodiversity and ecosystem function) was insensitive to edge lengths (i.e., observed relationship was not different than when edge lengths were randomized), and sensitive to the topology (i.e., observed values significantly different than when tip labels or internal nodes were randomized).

The simplest way to alter a phylogenetic tree, after swapping tip labels, is to change the edge lengths. We may want to stretch more recent edges if we want to test whether recent rapid evolutionary change is responsible for some pattern. Conversely, we may want to stretch deeper edges if we want to test if recent evolutionary change has stagnated and lineages rapidly diverged early on in their history (Simpson 1944, Gould and Eldredge 1977). We discussed some of these transformations in chapter 2, and we will return to these hypotheses later. In addition, there are various ways to randomize edge lengths. The first is to just assign random values to the edges—for example, pulling edge lengths from some distribution, such as the log-normal distribution. If we do this, we might want to standardize the overall tree depth so that we are not varying too many aspects of the tree simultaneously (relative edges lengths versus summed PD). Let's use the Jasper Ridge tree that we used in chapter 3 to do this. We will create two trees based on a single randomization. First we can simply randomize the existing edge lengths, which creates a nonultrametric tree, and then we use a rate-smoothing routine to make the tree ultrametric. For our purposes here, we will use the simple penalized likelihood approach (Sanderson 2002) using the `chronopl()` function in *ape* (see chapter 2 for more detail):

```
j.tree2<-j.tree
j.tree2$edge.length<-sample(j.tree$edge.length)
j.tree3<-chronopl(j.tree2,1,age.max=max(vcv(j.tree)))
```

TABLE 4.4.
Possible ways to alter phylogenetic trees to assess nonrandomness of community patterns.

Tree manipulation	Rationale	Fixed
1a. Branch lengths— random	Assess whether distances among species influence community patterns.	Topology
1b. Branch lengths— evolutionary rates	Assess whether changes in evolutionary rates influence community patterns.	Topology
2. Node position	Assess whether topology. influences community patterns	Distances
3. Branch lengths and nodes	Assess whether observed tree results in a particular community pattern relative to random trees.	None

When we use `chronopl()`, we need to specify the maximum node age, and here we want to ensure that it is the same as on our original tree to better allow comparisons between topologies. In the variance-covariance matrix representing the ultrametric tree, the diagonal values represent the tip to root distance, and are all the same. This distance is also the maximum distance in the matrix, so we just extract this maximum value from the matrix. We now have two trees with randomized edge lengths, one that is nonultrametric and one that is ultrametric (fig. 4.14) and we need to decide which to use. The answer is not clear. Some measures (e.g., MPD, MNND) work just fine on the nonultrametric tree. And since this tree has all the same edge lengths, it sums to the same PD value as the original (1508.4), while the ultrametric version has a different PD value (2312.4). This means that the average MPD will be larger in the ultrametric version, and so comparing the original observed MPD with this random distribution will be inherently biased. It is worth noting that it would be possible to rescale the ultrametric tree such that the total PD is summed to the original PD value, if this is important; however, there are other metrics that might be further biased by additional branch length transformations.

Now let's generate a tree with edge lengths pulled from a log normal distribution. For our purposes here we will only consider the ultrametric version scaled to the same tip-to-root distance as the original tree:

```
j.tree4<-j.tree
j.tree4$edge.length<-rlnorm(length(j.tree$edge.length))
j.tree4<-chronopl(j.tree4,1,age.max=max(vcv(j.tree)))
```

Now we can assess these two approaches on the significance of a particular community MPD value. We will stick with our previous example and examine "clump2b" from the *phylocom* data. The *phylocom* phylogeny is not helpful for this example since all of its internal branches are set to a length of 1, and therefore randomizing isn't meaningful. We'll use a new tree with branch lengths pulled from a lognormal distribution as described above. Because each time we generate the tree branch lengths will be different, we include this tree with the chapter files found at http://press.princeton.edu/titles/10775.html (the specific file is phylocom_tree_newDist.txt; shown in fig. 4.15). Just to make sure our new tree doesn't change the general pattern in the data, we can rerun the `ses.mpd` function with the tip-swapping randomization, and we can see that "clump2b" is still significantly underdispersed ($z = -4.04$, $P = 0.004$); we could also rate-smooth the tree so that it is ultrametric,

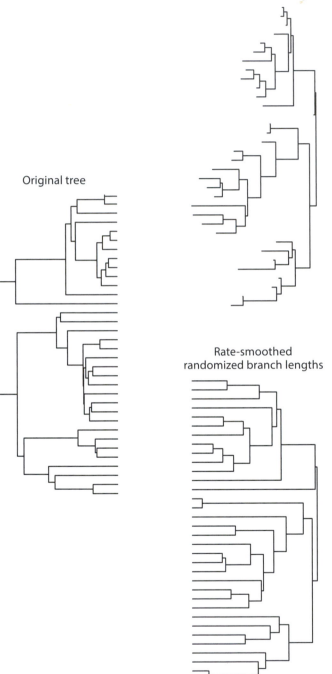

Figure 4.14. The output of a single randomization of the original Jasper Ridge tree is on the *left*. The randomized tree is nonultrametric (*top right*), and we can perform a rate-smoothing routine to make it ultrametric. The rate-smoothed tree looks similar to the original, but there are key differences in the branch lengths separating species.

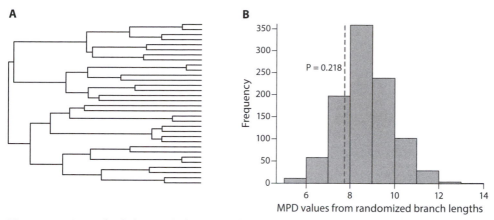

Figure 4.15. A sample phylocom phylogeny produced with edge lengths randomly sampled, and here rate smoothed for visualization (*A*). The observed MPD is not significantly different than the randomized MPD values (*B*).

but we will not do that here. The `ses.mpd` function does not implement the phylogenetic randomizations, so we will need to write the code for this:

```
tr.r<-read.tree("...phylocom_tree_newDist.txt")

mpd.obs<-mpd(phylocom$sample,cophenetic(tr.r))[3]

mpd.out.r<-NULL

for (i in 1:999){
  tr.tmp<-tr.r
  tr.tmp$edge.length<-sample(tr.r$edge.length)
  mpd.out.r[i]<-mpd(phylocom$sample,cophenetic(tr.tmp))[3]
}
```

Let's calculate the probability of the observed MPD:

```
obs.rank<-rank(c(mpd.obs,mpd.out.r))[1]
coef.obs.p <-obs.rank/(length(mpd.out.r)+1)
if (coef.obs.p > 0.5) out.p <-(1-coef.obs.p)*2
if (coef.obs.p <= 0.5) out.p <-coef.obs.p*2
```

As we can see in figure 4.15B, our observed MPD is not significantly different than the null values (*P* = 0.218). What this means is that the edge lengths in the phylogeny are not that important in generating the significant clustering of community members detected when we randomize the community membership. That is, topological patterns of relatedness are far more important than actual edge lengths. Here is the code for creating figure 4.15:

```
par(mfrow=c(1,2))
plot(tr.r,show.tip.label=FALSE)

hist(mpd.out.r,main=NULL,xlab="MPD values from randomized
     branch lengths")
abline(v=mpd.obs,lwd=2,lty="dashed",col="grey65")
text(7.9,250,"P = 0.218",col="grey65",pos=2)
```

The second major approach is to rescale edges in a way that alters their relative edge lengths near the root versus the tip; in essence, this is stretching deep branches and scrunching the tips or the converse. These are not randomizations, but they can be used to provide a distribution of values that can aid in developing inferences about how phylogenetic information influences ecological patterns. We can do this transformation manually by performing nonlinear transformations on the off-diagonals of the variance-covariance matrix, or use the `rescale()` function in the *geiger* package we used in chapter 2. As a reminder, this function performs a number of transformations of the phylogeny, and here we will again use "Pagel's delta (δ)" (see chapter 2) that represents a time dependent model of trait evolution (Pagel 1999). Recall that when δ is small (e.g., less than 1), recent evolution has been slow relative to deep branches; and when δ is large, then recent evolution has been more rapid than in the past (fig. 2.15). We can assess the influence of these evolutionary rates on our ability to detect community patterns (e.g., MPD). It is important to note that this analysis is not a randomization that produces a null alternative, but rather assesses the influence of a particular aspect of trees (here, tree depth).

Again, we can assess the significance of our observed MPD (i.e., $\delta = 1$) against those calculated from trees transformed by Pagel's δ. For this we will create transformed trees (see also chapter 2) and then run `ses.mpd()`, and ask how sensitive the significance found for the observed MPD using tip-label swapping is to the transformation of branch lengths. Here is the code to do this:

```
delta<-seq(0.1,10,0.1)

mpd.out.d<-NULL
p.out.d<-NULL

for (i in 1:length(delta)){
    tr.tmp<-tr.r
    tr.tmp<-rescale(tr.r,"delta",delta[i])
    mpd.out.d[i]<-mpd(phylocom$sample,cophenetic(tr.tmp))[3]
    p.out.d[i]<-ses.mpd(phylocom$sample,cophenetic
            (tr.tmp))$mpd.obs.p[3]

}
```

As we can see in figure 4.16A, increasing δ results in greater MPD values. We expect this because a high δ means the phylogeny is more starlike with longer distances between species. However, despite the change in absolute MPD values, the detection of the significant underdispersion is relatively insensitive to a range of δ values, as indicated by the flat *P*-value line at the bottom of the valley in figure 4.16B. We log transform the X-axis in figure 4.16, because Pagel's δ results in a nonlinear transformation of branch lengths. Again, we would conclude that the significance observed in the standard `ses.mpd()` test is relatively insensitive to the branch lengths and the model of evolution producing them.

Here is the code for plotting figure 4.16:

```
par(mfrow=c(1,2))
plot(log(delta+1),mpd.out.d,type="l",xlab="Log(Pagel's delta
        +1)",ylab="MPD",lwd=2)
points(log(2),mpd.obs,cex=2)
points(log(2),mpd.obs,cex=2.5)
```

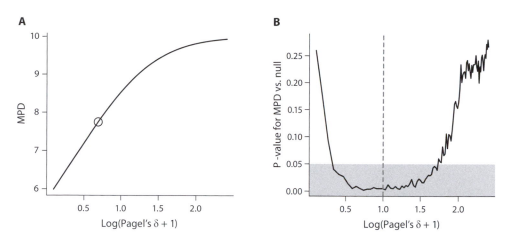

Figure 4.16. The change in MPD values with increasing Pagel's δ values (*A*). Detecting significantly clustered communities occurs at intermediate δ values (*B*).

```
plot(log(delta+1),p.out.d,type="l",xlab="Log(Pagel's delta
      +1)",ylab="P-value for MPD vs. null",lwd=2)
abline(v=1,lwd=2,lty="dashed",col="grey65")
```

In table 4.4 we suggested that there were three broad possibilities, and we have covered the first—altering edge lengths. The second—changing elements of the topology but not distances—is possible, but at the time of writing this book, we are not aware of a straightforward way to accomplish this. The final approach would be to use a randomized tree—for example, by using the `rcoal()` function and specifying the number of tips as the number of species in our data. We would then perform randomizations using `ses.mpd()`, much like we did for the previous Pagel's δ analysis. However, we would need to be cautious in our interpretation, since average MPD values would be different, and inferring mechanisms will be inherently difficult.

4.3.4. Other Randomizations and Inference Issues

There is always room to try new randomization tests, especially if a particular mechanism or hypothesis necessitates it. Building a null model should be done carefully, making sure that the inference consequences are well understood. Here are some examples of other approaches to building null assemblages:

- We can use a process-based model to construct communities. Constructing neutral communities (Hubbell 2001) would be one way to exclude a specific mechanism (interspecific competition) while returning others (dispersal limitation, carrying capacity). We could set the carrying capacity of communities to the observed abundances in those communities.
- We may realize that there are multiple assembly rules in play, and we wish to tease them apart. Our species may be divisible into guilds or functional groups, and we may want randomization strategies that maintain guilds, while allowing for random assembly within them (Wilson 1987).

- Species may occur in different communities because of known environmental affinities and perhaps we want to bias our randomizations to account for this. For example, if we have forest and grasslands in our data, it may not be meaningful for our hypothesis to have a randomization that allows forest trees in the grassland. One general method would be to assign probabilities for species inclusion based on known environmental affinities (Peres-Neto et al. 2001).

Regardless of the particular randomization approach used, we need to be careful about limits to inference and what mechanisms we evoke (Mouquet et al. 2012). Randomizing observational data is not the same thing as experimentally manipulating a set of factors with mechanistic underpinnings.

There are concerns that our randomization tests may obfuscate certain mechanisms, or that multiple mechanisms may produce similar patterns (Kraft et al. 2007, Hardy 2008, Kembel 2009, De Bello 2012). For example, Kembel (2009) showed with simulations that dispersal assembly may produce similar patterns to pure niche partitioning. Further, as suggested by Mayfield and Levine (2010), we might observe phylogenetic underdispersion with environmental filtering, as well as with competitive or fitness inequalities among clades. In their example, if particular species of plants are superior competitors for light (i.e., they grow taller), and plant height is phylogenetically conserved (see chapter 5) they may exclude species from other clades, thus decreasing average pairwise distances (see chapter 3 for further discussion on this).

4.4. TAKING THE POOL SERIOUSLY

It is often underappreciated just how important the species pool is to community phylogenetic analyses. Mean distances from a null distribution will be directly influenced by the topology of the species pool phylogeny (Cadotte 2014), and the power of any test will be influenced by the size and depth of the species pool (Cavender-Bares et al. 2006, Kraft et al. 2007, Lessard, Borregaard, et al. 2012). Let's look at an example of how the pool can affect significance testing, before outlining the decisions that need to be made.

The taxon labels in black in figure 4.17 represent the species present in a hypothetical community, and the gray edges show a distantly related clade that we will remove. First, here is the code to create figure 4.17:

```
tr.p<-read.tree("...Randomtree_pool.txt")
##for coloring tip labels in our community
com<-c("t1","t22","t24","t29","t18","t6","t11","t14")
cols<-rep("grey75",length(tr.p$tip.label))
cols[match(com,tr.p$tip.label)]<-"black"
plot(tr.p,cex=0.7,edge.color=c("grey55",rep("black",48),rep
    ("grey55",9)),edge.width=c(1.5,rep(2.5,48),rep(1.5,9)
    ),tip.color=cols)
```

Using the drop.tip() function from *ape*, we will create a new tree without the gray clade:

```
take.out<-c("t19","t26","t13","t12","t17")
sub.tr.p<-drop.tip(tr.p,take.out)
```

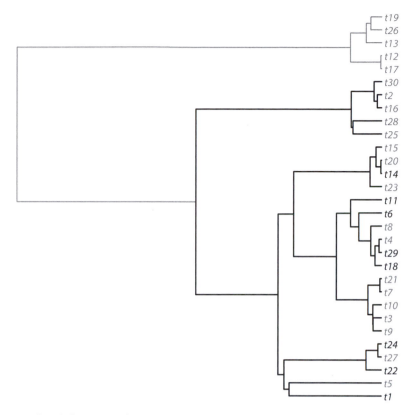

Figure 4.17. The phylogeny used in the species pool example. Edges in *black* show the subtree that contains the species (names in black) found in a community. The *gray* edges show the distantly related clade removed in the reanalysis.

Now we need to create the community data matrix:

```
com.mat<-matrix(0,ncol=length(tr.p$tip.label),nrow=1)
colnames(com.mat)<-tr.p$tip.label
rownames(com.mat)<-"Com1"
com.mat[1,match(com,colnames(com.mat))]<-1
```

We know how to run `ses.mpd()` from our previous examples and if we run it twice—once with the full tree and then with subset tree—we see a change in the significance test. The analysis with the full tree was significantly clustered ($z = -2.348$, $P = 0.018$), but using just the subset tree we find no significant community phylogenetic pattern ($z = -1.672$, $P = 0.082$). Thus, including a distantly related clade that is not represented in the local community can bias phylogenetic tests by increasing the null MPD values (Cadotte 2014).

This example highlights the importance of considering taxonomic scale, or the depth of the phylogeny (Cavender-Bares et al. 2006). Analyses could be limited to species from a single family, order, and so forth. What taxonomic scale should then be used for analysis? This is not a question easily answered. If we were performing an analysis on mammal assemblages in the Piedmont forests of the eastern United States, our data sets might include opossums along with a large number of other small and medium-sized placental mammals.

Being a marsupial, we know that opossums and the placental mammals in our piedmont assemblages diverged from a common ancestor 145 million years ago. This extremely long branch means that phylogenetic measures or null models will be highly skewed. Any local assemblage without opossums will likely appear significantly clustered. This underdispersion will not be the product of a specific ecological mechanism (e.g., environmental filtering) that we might be inclined to infer—but rather, the happenstance occurrence of a single species. This same problem exists for other taxa as well. Ferns and gymnosperms would have a similar effect on communities dominated by angiosperms.

There are a number of other issues beyond taxonomic scale that can similarly bias phylogenetic tests or inferences. Generally, if we are including disparate types of species (e.g., a predator with herbivores, or a tree with herbaceous plants) into a single analysis, then there may be artifacts present. In some cases, we may be including species that have no possibility of interacting, especially if our habitat scale is set too course, or if we include species with vastly different body sizes (e.g., elephants and rotifers) or that use completely different subhabitats (e.g., moles and monkeys). If our scale is too course and includes habitats containing species that do not interact, competition should not be used as an alternative hypothesis, which then only leaves us to test whether environmental filtering assembles communities in different habitats. By including multiple trophic levels, we might introduce phylogenetically clustered groups (e.g., predators) that interact with other species in ways (predation) for which we do not completely understand; that is, we do not have a clear hypothesis for the effect on community phylogenetic patterns (but see Cavender-Bares et al. 2006). That said, there might be good reasons to study phylogenetic patterns of prey choice and its impact, although ontogenetic shifts may obfuscate mechanisms and patterns across habitats; such shifts may similarly affect interpretion of species interactions.

Finally, biogeographic origin is an underappreciated issue in determining the appropriate pool. Frequently, our communities include non-native species, and we need to consider whether it is sensible to assume a single evolutionary model to explain potential differences or interactions among taxa. Species with much of their recent evolution intertwined with other local species may have coevolved in nonrandom ways that maximize local coexistence, but a pool of non-native species may be random with respect to local strategies (Cadotte and Jin 2014). One way to deal with this is to analyze native and non-natives separately to see if they show similar ecophylogenetic patterns (Cadotte, Borer, et al. 2010).

Another important issue is how we compile the list of species to go in the phylogeny. There are two main approaches for assembling the species pool: (1) from observed communities, and (2) from regional lists. Assembling from our observed communities provides us with a list of species we know can occur together. However, these communities may have been assembled through a strong structuring mechanism, and absent species may provide a signal of that mechanism. For example, if our community samples are all taken from shallow fast-moving stream reaches, our pool phylogeny will not include those species that occur in the same streams, but at slower, deeper reaches. The second approach, compiling from a regional species list can include species that would never normally occur in the habitat of interest (e.g., habitat scale), and so we need to carefully consider species inclusion and resulting inferences.

While excluding various species seems like the obvious way forward, this also needs to be carefully considered, as there could likely be repercussions. For example, we may decide to exclude gymnosperms from our analysis because of taxonomic scale concerns. However, the pine tree we removed may be a dominant species in our communities and certainly competes with angiosperms for light, nutrients, and water. Only by carefully thinking about the pool and running multiple analyses with and without certain groups can we hope to understand the impact that the pool has on our analyses.

4.5. CONCLUSION

Developing null expectations for community phylogenetic patterns from randomization procedures is the cornerstone for many ecophylogenetic analyses. While many studies frequently employ randomizations, there appears to be much less consideration of how communities and phylogenies should be randomized, and what specific randomizations mean for inference. We advocate an approach that explicitly considers multiple randomizations to better understand what aspects of community and phylogenetic structure determine observed patterns.

Some randomizations completely remove all structure, while others retain certain elements of the data (e.g., community richness, phylogenetic topology, etc.). This chapter considers various options that are currently implemented, but also highlights new approaches that also transform edge lengths on the phylogeny; these new approaches may provide greater insights into the mechanisms that structure communities.

Finally, we list a number of important considerations for constructing the species pool to be used in the randomization tests. The species that are included in the pool can have profoundly important consequences for the power to detect real patterns; conversely, species wrongly included can indicate nonrandom patterns when none exist. We advocate an approach that compares species pools, or at least links the pool to plausible ecological mechanisms.

CHAPTER 5

∞∞∞

Detecting Patterns of Trait Evolution

In every conceivable manner, the family is link to
our past, bridge to our future.
—*Alex Haley*

Throughout the chapters in this book we have assumed a relationship between the phylogenetic distance among species and their ecological similarities. As we have discussed previously, the general tenet is that more closely related species are assumed to be more ecologically similar, and thus have similar niche dimensions. This convenient presupposition hides a much more complex reality, and the relationship between phylogenetic distance and ecological difference hinges upon the mode of evolution of the characters under examination (Gittleman et al. 1996). For example, a trait such as body size in mammals might follow a more or less neutral model of evolution; although we might still expect upper and lower bounds to size limits, suggesting some evolutionary constraints to size. In contrast, flowering phenology, such as the time of first flower, might be a very labile trait, more closely tied to environmental change than intrinsic physiological constraints (Davies et al. 2013). Because the niche spans many ecological dimensions (Hutchinson 1959) and is shaped by many traits both independent and interacting, a relationship between phylogenetic distance and niche overlap might seem reasonable. Phylogeny allows us to integrate many traits, each of which might have their idiosyncratic evolutionary trajectory, but when averaged the mode of evolution might nonetheless resemble a neutral model. However, if just one or a few traits determine competitive strength or are important for accessing a limited resource, the mode of trait evolution might influence the shape of the relationship between phylogenetic distances and ecological differences. Fortunately, new methods allow us to search and contrast alternative models of evolution, and then transform the branches of the tree (see also chapter 2) to match to the best model. A first question to ask is whether the traits in question are phylogenetically patterned.

Although Linnaean classification preceded the development of evolutionary theory, Linnaeus was aware that some traits were more variable than others. His focus on the reproductive organs of plants, which might have seemed eccentric at the time, was largely pragmatic because he recognized that these traits tended to be less variable than characteristics of leaves or stems, and thus provided a more reliable feature for his classification system. Assumptions of phylogenetic trait conservatism underpin much of the phylogenetic comparative method and community phylogenetics. Predictions of phylogenetic clustering from filtering by the environment are predicated on the assumption that the climatic niche is phylogenetically conserved; whereas phylogenetic overdispersion from competitive displacement assumes phylogenetic conservatism in resource use (see chapter 3). We might, therefore, wish to estimate the strength of conservatism before attempting to infer the ecological processes structuring communities from their phylogenetic interrelationships.

5.1. PHYLOGENETIC SIGNAL

We can broadly define phylogenetic signal as the strength of the covariation in the differences in measured traits among taxa with the phylogenetic distances separating them (Blomberg et al. 2003). A strong phylogenetic signal would suggest that close relatives share similar traits (or trait values for continuous traits), while more distant relatives are less similar to each other. Traditionally, we might consider traits as morphological features, inherited with modification from a common ancestor, and similarity among individuals would thus reflect shared genotypes. However, traits can be more broadly defined as the expressed phenotype of an individual, a product of the interaction between genes and the environment. In the latter case, individuals might share similar traits because they share similar genes and/or environments. Phylogenetic conservatism in some species' features, such as the climate niche, might arise via inheritance of genetically conserved environmental tolerances or via phylogenetic conservatism in species geography. Phylogenetic conservatism in species geography might arise because, for example, the history of speciation is reflected in species' current distributions, as is most obviously the case in endemic island radiations (Wiens et al. 2010). In this latter example, there is no mechanistic underpinning between climate and species' distributions, but all species in the clade will likely share similar climate niches.

Traits can demonstrate a weak signal if they are predominantly environmentally determined. For example, the day of year on which a plant flowers might be a product of the amount of sunlight the plant has received over the preceding days or weeks. Two genetically identical plants might then flower at very different times of the year depending on their geographic location and local climate. This plastic response is not just noise to be avoided but can be useful to researchers wishing to track environmental change. For example, the Spring Indices that map the start of spring across of the northern hemisphere are derived from a phenological model of flowering times for cloned lilacs (Schwartz 1990). However, a plastic response in flowering time might still reflect underlying phylogenetic conservatism in response to phenological cues (Davies et al. 2013). A weak signal might also reflect strong directional selection pressure on phenotypes, resulting in evolutionary convergence (homoplasy), as in the lotus and water lilies discussed in chapter 1. With homoplasy, species are more similar to each other than predicted by their phylogenetic distances. It is important to recognize that the process of evolution itself can lead to weak phylogenetic signal in traits, and so a weak signal does not necessarily mean a lack of evolution. We typically measure the strength of the phylogenetic signal relative to some neutral model of evolution, in which expected variance increases in proportion to time since divergence. In such models the expected trait values have a normal distribution with a mean offset of zero.

5.1.1. Continuous Data

For continuous traits (i.e., data that can have any numeric value along a continuum), we commonly assume a Brownian motion (BM) model of trait change, in which traits diverge independently over time in a manner analogous to a random walk. The BM model is attractive because it allows many solutions to be derived analytically, without the need for simulations, and thus it is highly tractable. Under BM change in trait X over time t can be simply represented as:

$$dX(t) = \sigma dB(t),$$

where $dB(t)$ is a normal random variable with a mean of zero and variance $\sigma^2 dt$; thus the rate of evolution is simply given as σ^2, which we discuss further below. Many R libraries allow us to simulate BM trait evolution along the branches of a phylogenetic tree; here we use the *geiger* library introduced earlier to simulate a trait (X) assuming a σ^2 of 0.2 on a simulated tree of size $n = 64$ using the `rcoal()` function from the *ape* library (see chapter 2 for *ape* basics). This generates an ultrametric tree, such that tips line up in the present, by randomly clustering tips. Here is the code:

```
tree<-rcoal(64)
X<-rTraitCont(tree, model = "BM", sigma = 0.2)
```

Because we have simulated X under a BM model, we might expect it to demonstrate a strong phylogenetic signal. There are two commonly used indices of phylogenetic signal for continuous variables—Pagel's λ (Pagel 1999) and Blomberg's K-statistic (Blomberg et al. 2003). Both indices evaluate strength of the signal relative to the BM model, and are conveniently bounded at zero (representing no signal) with values of one equal to expectations from BM. In some cases, both λ and K can return values >1; for K this represents a model in which species are more similar in their trait values than predicted from BM, while λ values >1 are undefined. Pagel's λ can be estimated in *geiger*:

```
X.Lambda<-fitContinuous(tree,X,model="lambda")
```

Here we are provided with the inferred estimate of λ along with the model log likelihood (lnL). We can therefore assess whether our measured trait X shows a greater phylogenetic signal than expected by comparing the likelihood of the model to the fit of X, to a star phylogeny. First we need to convert the bifurcating coalescent tree to a star phylogeny. There are various ways we could do this, but for consistency, we use here the `rescale()` function in *geiger*:

```
star.tree<-rescale(tree,"lambda", 0)
```

If you examine this tree closely (`summary(tree)`) you might notice that the tree is still fully resolved, and transforming the tree using `lambda = 0` simply sets internal branch lengths to zero, which is sufficient for our purposes here. Next, we estimate the fit of X on the star phylogeny:

```
X.Lambda.star<-fitContinuous(star.tree,X,model="BM")
```

Then we can use a simple likelihood-ratio test to compare model fits with significance evaluated assuming a chi square distribution on one degree of freedom:

```
LR<-2*(X.Lambda$opt$lnL-X.Lambda.star$opt$lnL)
pchisq(LR,df=1,lower.tail=F)
```

As we would expect, we found a highly significant difference between model fits. That is, the fit of X to the true tree is significantly better than the fit of X to the star phylogeny, which contains no information on evolutionary relationships. Therefore, we can infer that our trait demonstrates significant phylogenetic structure. We might also be interested in whether X differs from a BM model, and following the same procedure as above we can transform our phylogeny using the maximum likelihood estimate of λ, and compare the likelihood of this model to that on the untransformed tree, which, in this case represents expectations from BM:

```
transform.tree<-rescale(tree, "lambda",
    X.Lambda$opt$lambda)
X.Lambda.transform <-fitContinuous(transform.
    tree,X,model="BM")
LR<-2*(X.Lambda$opt$lnL-X.Lambda.transform$opt$lnL)
pchisq(LR,df=1,lower.tail=F)
```

This time the P-value approaches one, indicating that the fit of X to the phylogeny is not significantly different from the BM expectation, which is reassuring since we simulated the trait using a BM model!

How does Blomberg's K compare? We can calculate the K-statistic easily in the *picante* library:

```
Kcalc(X, tree)
```

Note that by default this function checks that the order of trait values match the order of tips in the tree; however, it is good practice to confirm this by reordering our vector of traits by `tip.label`:

```
ordered.X<-X[tree$tip.label]
Kcalc(ordered.X, tree)
```

Reassuringly, our K-values are unchanged. `Kcalc()` can also be called on a column in a data frame, but again it is good practice to check that the order of values matches that of tip labels in the tree.

You might notice that K for X is often less than (and sometimes greater than) one (try running the code a few times). Should we interpret this as evidence for lack of phylogenetic signal? As for Pagel's λ, we can compare the observed K to a null expectation based on absence of signal. As we mentioned in chapter 4, we can do this by randomly resampling the data, and recalculating K, here using 99 replicates (you might want to run 999 or more to generate a more stable null distribution, but this may take some time):

```
rand.K<-NULL
for (i in 1:99){
rand.K[i]<-Kcalc(sample(X), tree, checkdata=FALSE)[1]
}
```

In the randomization you will note that we specify `checkdata=FALSE` to break the link between taxa names in our trait data frame and tree tip labels. We can then plot the distribution of null values using the histogram function and examine whether the estimate of the observed K-value falls outside the 95 percentile of the null distribution (assuming a one-tailed test and significance at $\alpha = 0.05$; see chapter 4 for further discussion of null models):

```
hist(rand.K)
abline(v = Kcalc(X, tree), col = "gray", lwd = 2)
```

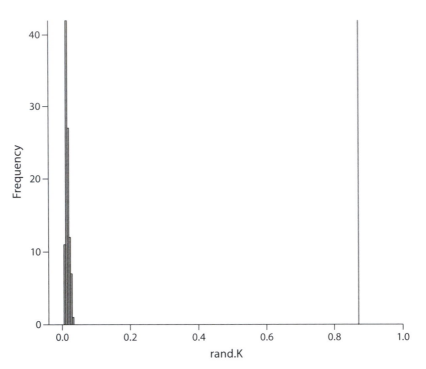

Figure 5.1. Frequency distribution of K values generated from randomized data, in comparison to observed K from a trait simulated on the phylogeny by Brownian motion (vertical gray line).

For this example the histogram displays perfectly, but we cannot see where the empirical value of K falls—this is because the observed K is so far outside the distribution of our random expectations that the x-axis of the histogram does not stretch far enough for us to see it (fig. 5.1). However, we can force the plot window to extend the *x*-axis of the histogram between zero and one, which bounds most values of K, and we can replot the data. (However, if the observed value is still not apparent—that is, it is larger than 1, then we can extend the maximum value on the *x*-axis by setting `xlim=c(0, Kcalc(X, tree)+0.3)`:

```
hist(rand.K, xlim=c(0,1.5))
abline(v = Kcalc(X, tree), col = "gray", lwd = 2)
```

The observed K-value is plotted as a vertical gray line (fig. 5.1), and it is now obvious that the observed K falls far outside of our null distribution; we can be confident in concluding that phylogenetic structure in *X* is significantly greater than predicted by chance (i.e., randomly assigning trait values to the tips of the tree). This randomization procedure provides a robust statistical test to evaluate phylogenetic structure, but it can be time consuming because estimating K is computationally demanding. This is especially true when evaluating significance in K across large data sets containing many hundreds of species, or, as another example, across multiple phylogenetic topologies using a sample of the posterior distribution of trees generated from Bayesian analyses. Fortunately, we can use a shortcut implemented in the *picante* library that compares the expected variance in phylogenetically independent contrasts (PIC). Felsenstein (1985) introduced PICs to control for the evolutionary nonindependence of species in comparative analyses. The method has been discussed extensively in the literature, and remains a mainstay of comparative biology. With PIC, differences between mean species values are calculated across each node in the phylogeny, weighted by the expected variance assuming a BM model of evolution; this is because we assume that the expected variance would be greater across nodes deeper in the tree due to greater time for evolutionary divergence. Independent contrasts were first introduced to evaluate evolutionary correlations between species traits—for example, brain size and body size in primates—but here we can use them to evaluate significance of phylogenetic structure in the data. If species' traits follow a BM model of evolution, the weighted differences calculated across the set of nodes in the phylogenetic tree should be normally distributed, and the variance in contrasts should be low. However, if traits are distributed randomly among tips, variance in contrasts will be high, because there is no relationship between phylogenetic distance and trait differences. We can thus evaluate significance in the phylogenetic structure of trait distributions by comparing observed variance in PICs to a null distribution generated again by shuffling trait values among tips. We can do this manually, as we did for Blomberg's K:

```
X.PIC<-var(pic(X, tree, scaled = TRUE))

rand.PIC<-NULL
for (i in 1:100){
rand.PIC[i]<-var(pic(sample(as.vector(X)), tree))
}

hist(rand.PIC,xlim=c(0,10))
abline(v = X.PIC, col = "gray", lwd = 2)
```

Here we can see that our observed estimate of variance in PICs is again far outside of the null distribution (fig. 5.2), confirming results from randomization of K that the data show significant phylogenetic structure. The `phylosignal()` function in *picante* provides us with a simple wrapper to calculate Blomberg's K and significance from PICs, and requires much less computational time than the randomizations using `Kcalc()`:

```
phylosignal(X, tree, reps = 999, checkdata=TRUE)
```

Here we specify the number of replicates for estimating significance, and again we can specify whether names in the data file match to tree `tip.label`.

Both Pagel's λ and Blomberg's K should be independent of tree size, although our power to detect significant structure might be reduced for small trees (Blomberg et al. 2003). However, incompletely resolved trees might bias estimates. At the extreme, a star phylogeny will always return K = 1, even for random data:

```
Y<-runif(length(tree$tip.label))
names(Y)<-tree$tip.label
Kcalc(Y, star.tree)
```

Thus our estimates of phylogenetic signal might be inflated for poorly resolved trees. We need to be careful of this bias, as many approaches for generating large comprehensive phylogenetic trees simply add taxa as polytomies to the minimally inclusive node when additional data on phylogenetic placement is lacking (see chapter 2). One simple solution is to thin the tree, pruning taxa from the polytomies, leaving a completely bifurcating topology.

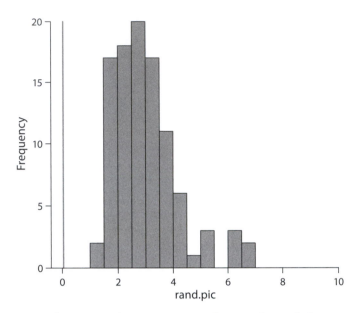

Figure 5.2. Frequency histogram of variance in PICs from randomized data, in comparison to observed K from a trait simulated on the phylogeny by Brownian motion.

For polytomies deep in the tree, thinning descendent lineages might remove a very large number of taxa; as a compromise, we might therefore restrict ourselves to thinning only terminal polytomies (which in any case tend to be more common). Nathan Kraft from the paper by Davies, Kraft, et al. (2012) provides us with a useful function:

```
thin_terminal_polytomies=function(phy, keep=1){

    phy$edge->edges
    length(phy$tip.label)->ntips
    edges[which(edges[,2]<=ntips),]->terminal_edges
    names(which(table(terminal_edges[,1])>2))->poly_list
    droplist<-NULL
    for(i in 1:length(poly_list)){
    terminal_edges[which(terminal_edges[,1]==poly_
        list[i]),2]->tips
      droplist<-c(droplist, sample(tips,
            size=(length(tips)-keep)))
    }
    drop.tip(phy, droplist)->phy_thin
    return(phy_thin)
}
```

First let us generate a tree containing a polytomy using the `read.tree()` function:

```
poly.tree<-read.tree(text="(t1:4,(t2:3,(t2:2,(t3:1, t4:1,
    t5:1):1) :1) :1);")
```

In this example, tips *t3*, *t4*, and *t5* form a polytomy. Now we can thin this tree using the function above:

```
thin.tree<-thin_terminal_polytomies(poly.tree, keep=1)
```

We would then estimate K on the thinned tree, as described above. For large trees with many polytomies, it would be sensible to randomly thin the tree multiple times, estimating K on each thinned topology.

5.1.2. Discrete Data

Equivalent models for estimating phylogenetic structure for discrete traits are also available. One approach uses parsimony to estimate the number of state changes along the branches of the tree, and the sum of edge lengths can then be compared to those estimated from a randomized matrix of trait data. This approach is commonly implemented in the Mesquite phylogenetic software (Maddison and Maddison 2015), and has been discussed elsewhere (e.g., see Miyamoto and Cracraft 1991). Because assumptions of parsimony are often violated, we focus here on maximum likelihood approaches. We can adopt a similar approach to continuous data, using Pagel's λ as described above, but substitute fitContinuous() with its discrete counterpart, fitDiscrete(). First, let us transform our continuous trait data into a binary variable, assigning values greater than the clade mean a value of 2, and values less than the clade mean a value of 1:

```
Z<-(X>mean(X)) +1
names(Z)<-names(X)
```

Now we can fit the model of evolution to our data:

```
discrete.model<-fitDiscrete(tree,Z,transform="lambda")
```

Here the value reported for optlambda gives our ML estimate of λ. Because we evolved our data under a Brownian motion model of evolution, we would expect our estimate of λ to approach 1. This can be checked by typing discrete.modeloptlambda. We can then evaluate significance as for continuous variables.

Blomberg's K is not suited to evaluating signal for discrete data, but we can use the D-statistic of Fritz and Purvis (Fritz and Purvis 2010). This implements a method similar to PICs to test for significant phylogenetic structure in discrete data, and is implemented in *caper*. First, we must generate a new data frame with a column of species names:

```
D<-cbind(as.data.frame(Z), names(Z))
colnames(D)<-c("Z", "taxa")
```

And then we calculate the D-statistic using the phylo.d() function, where the names. col specifies which column in the trait data frame contains taxon names that match to tip labels, while binvar specifies the column containing the binary data:

```
phylo.d(D, tree, names.col = taxa, binvar = Z)
```

Somewhat confusingly, here a value of 0 equates to a BM model, and a value of 1 to no signal. As for continuous traits, significance is assessed by shuffling trait values across taxa and *P*-values are presented to compare the fit to a BM model and a random null expectation.

The strength of phylogenetic signal in trait data may be of interest in its own right. For example, strong phylogenetic conservatism in the climate niche might help explain biogeographical gradients of species richness (Wiens and Donoghue 2004), and also the capacity to adapt to future, changing environments (Lavergne et al. 2013). Within the field of eco-phylogenetics, we are more interested in how accurately phylogeny might capture species differences; phylogenetic distances might be a better representation of species differences for traits exhibiting strong phylogenetic conservatism. However, many traits might not evolve in a Brownian fashion, yet are still strongly structured by phylogeny.

5.2. ALTERNATIVE MODELS OF TRAIT EVOLUTION

Brownian motion is the most commonly assumed model of evolution, at least for continuous traits. However, there are strong arguments to suggest that alternative evolutionary models might be more commonplace. It is now widely recognized that evolution can depart significantly from Brownian expectations, and that models of evolution may vary among traits, clades, and taxonomic ranks (Gittleman et al. 1996, Hansen 1997, Böhning-Gaese and Oberrath 1999, O'Meara et al. 2006, Cooper and Purvis 2010).

5.2.1. Evolution with Constraints

Perhaps the simplest deviation from BM is a model of constrained evolution, in which trait divergence follows a random-walk model with a central tendency, such that phenotypes are pulled back toward some optimal trait value (Hansen 1997). This model is sometimes referred to as the Ornstein-Uhlenbeck (OU) model, and might be conceptualized as stochastic evolution with a constraining force due to stabilizing selection (Hansen 1997). Following notations for BM, above, we can represent the OU model as follows:

$$dX(t) = \alpha \, [\theta - X(t)]dt + \sigma dB(t)$$

Here the additional term ($\alpha \, [\theta - X(t)]dt$) represents stabilizing selection, with α as the strength of selection pulling the trait value toward the optimum trait value, θ. Because trait space is typically bounded, we might generally expect the OU model to better characterize the mode of trait evolution. In terrestrial mammals, variation in body size spans many orders of magnitude, from just a few grams (e.g., bumblebee bat, *Craseonycteris thonglongyai*) to several tons (e.g., African elephant, *Loxodonta africana*). Sizes much smaller than two grams may be constrained by physical limits, such as the thermodynamics of the electron transport chain (Dobson and Headrick 1995), whereas biomechanical limits might explain why the largest terrestrial mammals never greatly exceeded the extant African elephant in size (Smith et al. 2010). Energetic and mechanical constraints were likely even more limiting for flying birds. Mammals were able to escape mechanical constraints to upper-body size limits in the marine environment, and here we witness the spectacular radiation of cetaceans, which includes the blue whale (*Balaenoptera musculus*), perhaps the largest animal ever to have lived. Similarly, flightless birds may have been able to escape the upper constraints to body size that limited their volant relatives. The ostrich (*Struthio camelus*) is the largest extant bird, and is almost an order of magnitude heavier than the heaviest flying species (the kori bustard, *Ardeotis kori*); and the now extinct but also terrestrial Striton's thunderbird, *Dromornis stirtoni*, may have attained a size close to 500 kg.

Much as we compared the fit of the BM model to the null model of no phylogenetic signal, we can contrast the fit of alternative evolutionary models using the fitContinuous() function in *geiger*, and specifying the alternative evolutionary model; here we fit the BM and OU models to the trait data we simulated under BM, above:

```
X.BM<-fitContinuous(tree,X,model="BM")
X.OU<-fitContinuous(tree,X,model="OU")
```

As before, we can compare model fits using likelihood ratio tests, noting that the OU model has two additional parameters, α and θ. However, fitContinuous() also provides us with other model fit statistics, and we can just as easily compare models using Akaike information criterion (AIC); for good practice, it is best to use the correction for small sample size (AICc). Following convention, we might consider a difference in AIC of >3 to suggest support for one model over the alternative, and a difference of >10 indicates overwhelming support for the model (Burnham and Anderson 2002); remember, we favor the model with the lower AIC:

```
X.BM$opt$aicc - X.OU$opt$aicc
```

In our example, in which we evolved traits under BM, we would not expect the OU model to be favored over BM. But let us now generate traits under an OU model and repeat the model comparison:

```
X<-rTraitCont(tree, model = "OU", sigma = 0.1, alpha = 1,
        theta = 0)
X.BM<-fitContinuous(tree,X,model="BM")
X.OU<-fitContinuous(tree,X,model="OU")
X.BM$opt$aicc - X.OU$opt$aicc
```

There is some stochasticity, because the underlying process of evolution that we are modeling is inherently noisy, but you might notice after running the code a few times that we still do not observe consistent support for the OU model, even though we know this is the "true" model. This illustrates an important limitation to the model comparison approach; because AIC penalizes parameter-rich models we will tend to favor the simpler model when data is noisy and signal is weak. In our simulated data, signal for the OU process may be weak because the rate of evolution (sigma) is relatively low, and the strength of stabilizing selection (alpha) is relatively weak, such that simulated traits might only rarely obtain values where the strength of stabilizing selection exerts a notable force. A simple exercise would be to incrementally increase alpha and sigma, and test whether the OU model is supported more strongly.

If we find that the BM model is not strongly supported, the phylogenetic distance between taxa might not do a good job of capturing expected differences in trait values. We can demonstrate the effect of the evolutionary model on *patristic distance* on the phylogeny by transforming the "edge.lengths" of the underlying phylogeny using the α parameter from the OU model, and plotting the two trees side by side (fig. 5.3):

BM tree   OU tree

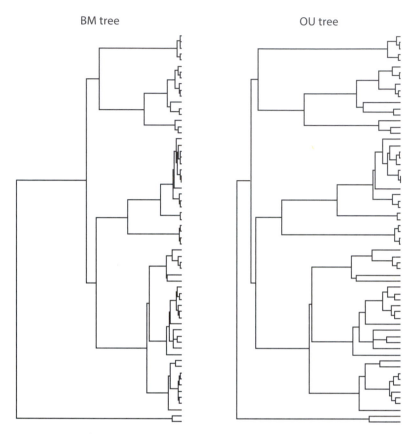

Figure 5.3. Comparison between two trees scaled to represent different evolutionary models of trait evolution: Brownian motion (BM) and the Ornstein-Uhlenbeck (OU) models of stabilizing selection.

```
OU.tree<-rescale(tree,model = "OU", X.OU$opt$alpha)
par(mfrow=c(1, 2))
plot(tree)
title("BM tree")
plot(OU.tree)
title("OU tree")
```

We can see clearly how the OU transformation disproportionally stretches branches toward the tips of the tree. Because many metrics of community structure, such as MPD and MNND (see chapter 3), are derived from pairwise distances between taxa on the phylogeny, they might not capture the underlying processes structuring species coexistence if traits deviate from strict BM assumptions (Kraft et al. 2007). One simple solution would be to use the OU transformed tree topology to estimate community phylogenetic structure.

5.2.1.1. Multirate and Multioptima Models

An extension of the OU model implemented in the *OUwie* library allows more complex evolutionary models that account for clade differences in evolutionary rates (σ^2), optima (θ), and strength of stabilizing selection (α). We can therefore specify a set of nested models varying in complexity:

BM1 – single rate BM
BMS – multiple rate BM
OU1 – single optima OU
OUM – multiple optima OU
OUMV – multiple optima OU with multiple rates
OUMVA – multiple optima OU with multiple rates and multiple alphas

To fit these more complex models, where parameter values vary by clade, we first have to assign each lineage on the phylogeny to a selective regime. For example, we might wish to assign regimes based upon particular ecological features, such as growth form or habit, depending on the hypothesis being investigated. One approach is to estimate ancestral character states (see below) and then assign nodes to the appropriate states. First, let us simulate a new discrete trait K using `rTraitDisc()` from the *ape* library:

```
K<-rTraitDisc(tree, states=c(1,0) , rate = 0.4)
```

Note that we have specified a binary 0:1 trait using `states`, and set the `rate` to 0.4, which is greater than the default value of 0.1; we do this here to ensure that the simulated trait values are variable enough to be interesting for our analysis. Check the values of K before moving on, and if all tips have identical states, simply try rerunning this line of code. Second, we can reconstruct the ancestral states using ancestral character estimation (`ace()`) in the *ape* library:

```
node.ace<-ace(K, tree, type = "discrete")$lik.anc
```

For discrete traits, the ancestral states are estimated using maximum likelihood (see below for further details). Now we have the scaled likelihoods of each ancestral state (`...$lik.anc`) at each node. You might have noticed that `rTraitDisc` actually has the option to return internal node estimates (`ancestor = TRUE`), but as this information will not be available for most empirical data sets, we will continue as if we are ignorant of the true ancestral states. For simplicity, we will assign nodes to the ancestral state with the largest likelihood:

```
my.regime<-node.ace[,1]>0.5
tree$node.labels<-my.regime
```

We can view the assignments of nodes to regimes by plotting them on the phylogeny (fig. 5.4):

```
select.reg<-NULL
select.reg[tree$node.label == 0] <-"gray20"
select.reg[tree$node.label == 1] <-"gray50"
plot(tree, show.tip.label = F)
nodelabels(pch=19, col=select.reg)
```

Third, we now construct a data frame with tip labels, regime, and our trait X, which we generated earlier in this chapter and which we wish to model:

```
q<-data.frame(names(K), K, X)
colnames(q)<-c("species", "regime", "trait")
```

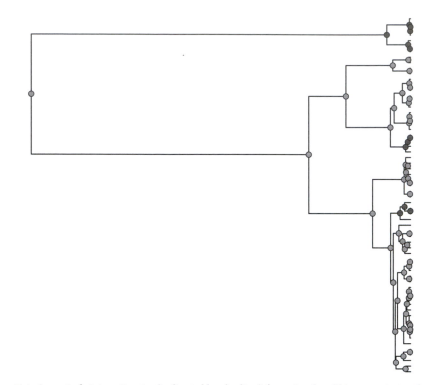

Figure 5.4. Ancestral state estimates (indicated by shading) for a simulated binary trait simulated and reconstructed using maximum likelihood (see text).

Fitting the alternative models is then straightforward:

```
BM1<-OUwie(tree,q,model="BM1")
BMS<-OUwie(tree,q,model="BMS")
OU1<-OUwie(tree,q,model="OU1")
OUM<-OUwie(tree,q,model="OUM")
OUMV<-OUwie(tree,q,model="OUMV")
OUMVA<-OUwie(tree,q,model="OUMVA")
```

We can then compare their fits using AICc, as described above. Here, we would expect the BM1 model to be favored, as we have no reason to expect model parameters to covary with the regime K, which we simulated independently from our trait data, X. Because the OUMV and OUMVA models are parameter rich, it is possible that the complexity of the model might be greater than the information contained within the data, resulting in poor parameter estimation.[1] *OUwie* will provide a warning message if the data are insufficient to fit a particular model, but in some cases the ML search might simply fail, depending

---

1. In the help section in OUwie it is suggested that an eigen decomposition of the Hessian matrix, containing second-order derivatives, may provide an indication of whether parameter estimates are reliable; if all the eigenvalues of the Hessian are positive, then the parameter estimates are considered reliable. It is also possible to identify offending parameters through examination of the eigenvectors. Further details can be found in the references included for this function.

on the tree shape and trait data supplied (you can evaluate alternative tree topologies and simulated data to further explore the performance of this function). One simple option is to stick with less complex models.

One additional variant of the BM model that also incorporates variable rates is Pagel's δ model, which we are familiar with from chapters 2 and 4. In this context, we can interpret the model as representing changes in evolutionary rates (σ) through time. For example, we might predict that clades undergoing initial stages of adaptive radiations, in which they are expanding into a new niche space, might show accelerated evolutionary rates toward the present, whereas evolutionary slowdowns might represent a process of niche filling (Phillimore and Price 2008, Rabosky and Lovette 2008). As a reminder, Pagel's δ model raises node depths to the power δ; $\delta = 1$ represents BM; $\delta > 1$ (speedup) disproportionately stretches branches toward the tips of the tree, such that evolution is concentrated at the tips of the tree; and $\delta < 1$ (slowdown) disproportionately stretches branches toward the root of the tree, such that evolution is concentrated early in the tree (see chapters 2 and 4 for more discussion on Pagel's δ). We can fit this model simply and compare it to a BM model, using similar syntax to the λ model.

There are, of course, other variations that we might want to consider; for example, the early burst (EB) model in fitContinuous() allows for rates to increase or decrease exponentially, as well as more complex models utilizing Bayesian approaches that identify locations of potential rate transition on the tree (Revell et al. 2012). Theoretically, it is possible to transform the underlying tree topology to fit to each of these models, although it should be recognized that such models are likely to be highly specific to the particular trait under consideration; therefore patristic distances in transformed trees may be less appropriate for capturing differences in independently evolving traits. However, there are certain cases where we might want to use the transformed tree structure—for example, if we are interested in evaluating evidence for character displacement in a particular trait (see chapter 6).

5.2.2. Speciational Models and White Noise

So far we have considered variants of the BM model. A third Pagel transformation, κ, which we also discussed in chapter 2, allows trait evolution to occur in bursts at speciation events, thus representing Gould's theory of punctuated equilibria (Gould and Eldredge 1977). In this model, the number of evolutionary splits (nodes) separating taxa may be a better indication of phenotypic differences between them than the sum of the branch lengths. Here, each branch of the tree is raised to the power κ, such that $\kappa = 1$ is BM, and $\kappa = 0$ is a pure speciation model (all branches are equal).

As before, we can visualize the model by transforming the underlying tree topology (fig. 5.5) here, illustrating extreme values of κ:

```
speciational<-rescale(tree, model = "kappa", 0)
BM<-rescale(tree, model = "kappa", 1)

par(mfrow=c(1, 2))
plot(speciational)
title("speciational")
plot(BM)
title("BM tree")
```

Speciational   BM tree

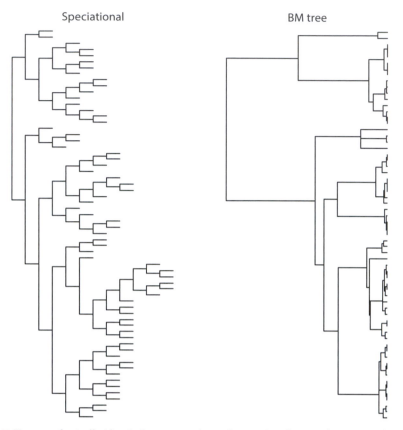

Figure 5.5. Two topologically identical trees transformed using Pagel's $\kappa = 0$ (speciational), and $\kappa = 1$ (Brownian motion).

Conveniently, we can evaluate the fit of κ within the same AIC framework as for δ and λ:

```
X.Kappa<-fitContinuous(tree,X,model="kappa", bounds =
        list(kappa = c(min = 0, max = 2)))
```

We can see clearly that that the κ model provides a poor fit to our data, as we might expect, given that we originally generated the data under BM.

Finally, it is worth introducing the white noise model, in which trait values are drawn from a normal distribution:

```
X.WN<-fitContinuous(tree,X,model="white")
```

Support for the white noise model would suggest there is no phylogenetic signal in the data. We might therefore compare the fit of a white noise model to that of BM or another explicit evolutionary model to evaluate phylogenetic signal in the data; support for a white noise model would indicate low phylogenetic signal (i.e., weak phylogenetic conservatism).

5.3. RECONSTRUCTING ANCESTRAL STATES

Community ecologists might appear to have little to gain from reconstructing the ancestral states of lineages over deep time, because the ecological interactions between co-occurring species is inarguably determined by their current traits, not those of their ancestors. But it is easy to forget that the traits we observe among species today are a product of a long evolutionary history. The theory of limiting similarity (Gause 1934), an extension of the competitive exclusion principle (Gause 1934, Abrams 1983), which has shaped ecological thinking for the better part of a century, states that two species competing for the same resources cannot coexist indefinitely. Thus, coexisting species must differ in a manner that allows them to exploit different resources (MacArthur and Levins 1967, Chesson 2000, Adler et al. 2007). If species are too similar, competition will result in selection for character displacement, and thus divergence in their resource use (Brown and Wilson 1956). These deceptively simple predictions have proven much more difficult to test in practice. One common approach has been to search for evidence of niche segregation and then infer evolutionary character displacement. In what has become one of the classic papers in ecology, Joseph Connell (1980) refers to such efforts as a call to the *ghost of competition past*. By providing information on ancestral states, phylogenetics allows us to revisit these questions within an explicitly evolutionary framework.

Here we cover some of the common models for reconstructing ancestral states for discrete and continuous data. This section is intended as an introduction to the topic, and focuses on applying the `ace()` function in the *ape* library; the interested reader is encouraged to explore the literature further.

5.3.1. Reconstructing Continuous Data

Let us generate a new tree and a new continuous trait, X:

```
tree<-rcoal(64)
X<-rTraitCont(tree, model = "BM", sigma = 0.1)
```

We then reconstruct the character states at the internal nodes on the tree:

```
anc.REML<-ace(X, tree, type ="continuous", method = "REML")
```

For `type = "continuous"`, the default model is BM, and the model is fitted by `method = "REML"` (Restricted Maximum Likelihood), which is recommended over ML (Maximum Likelihood). Ancestral values can be called by `...$ace`. For very large data sets, method = `"pic"`, which uses ancestral estimates from phylogenetic independent contrasts, might be preferable because it is highly tractable and provides a rapid solution and thus limits the computational time required to complete the analyses. Generally, estimates using REML, ML, and PIC should be similar and the key difference is that node estimates from PIC consider only descendent species values. Finally, `method = "GLS"`, in which character states are reconstructed using generalized least squares (Martins and Hansen 1997, Cunningham et al. 1998), is more flexible in that it allows the user to specify the correlation structure of the model (`corStruct`); however, it has been reported to be less stable than other methods. We can quickly compare model estimates by plotting reconstructed values on the internal nodes of the phylogeny (fig. 5.6):

REML reconstruction   ML reconstruction   PIC reconstruction   GLS reconstruction

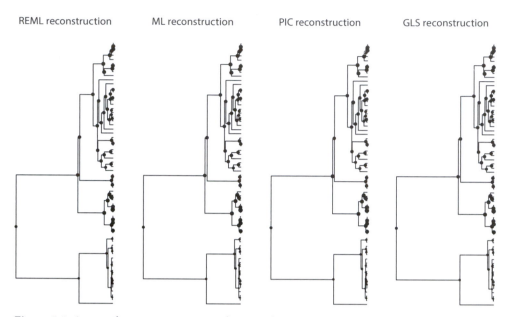

Figure 5.6. Ancestral state reconstruction for a single continuous trait using four different reconstruction methods (see text); symbol size is proportionate to reconstructed trait values.

```
anc.REML<-ace(X, tree, type ="continuous", method =
     "REML")$ace
anc.ML<-ace(X, tree, type ="continuous", method = "ML")$ace
anc.PIC<-ace(X, tree, type ="continuous", method =
     "pic")$ace
anc.GLS<-ace(X, tree, type ="continuous", method = "GLS",
     corStruct = corBrownian(1, tree))$ace

anc<-anc.REML + abs(min(anc.REML))
plot(tree, show.tip.label=F)
nodelabels(pch = 19, cex=(anc/max(anc)), col = "red")
title("REML reconstruction")

anc<-anc.ML + abs(min(anc.ML))
plot(tree, show.tip.label=F)
nodelabels(pch = 19, cex=(anc/max(anc)), col = "red")
title("ML reconstruction")

anc<-anc.PIC + abs(min(anc.PIC))
plot(tree, show.tip.label=F)
nodelabels(pch = 19, cex=(anc/max(anc)), col = "red")
title("PIC reconstruction")

anc<-anc.GLS + abs(min(anc.GLS))
plot(tree, show.tip.label=F)
nodelabels(pch = 19, cex=(anc/max(anc)), col = "red")
title("GLS reconstruction")
```

In figure 5.6 the symbol size is proportionate to reconstructed trait values; note that we have to first positivise values by adding the absolute value `abs()` of the minimum reconstructed value, before plotting. Reassuringly, we can observe that reconstructed values are similar for all methods. However, in our example we simulated the trait data under a BM model, matching assumptions of the model fit methods. We might expect methods to diverge more for empirical data, where strict BM assumptions might be violated. In such cases, it might be sensible to first evaluate the model of evolution, as described above, and then transform the tree appropriately before estimating ancestral states.

5.3.2. Reconstructing Discrete Data

Options for reconstructing ancestral states for discrete data (`type = "discrete"`) are more limited. Only maximum likelihood approaches (Pagel 1994) are implemented in `ace()`, but the syntax is very similar to that for continuous data. We briefly introduced the approach above when we assigned selection regimes to models of trait evolution; here we provide some more details. Let us simulate a new discrete trait K, as above, but this time with three states, (0/1/2):

```
K<-rTraitDisc(tree, k = 3, states=c(0,1,2) , rate = 0.4)
```

We then reconstruct the character states at the internal nodes on the tree (make sure you have a reasonable sample of all three states in your simulated traits):

```
anc.ER<-ace(K, tree, type = "discrete", model="ER")
```

You will notice that we now specify the model to be used in the reconstruction, which describes the underlying rate transition matrix between character states. We are provided with three standard options: ER (equal rates, a one parameter model in which transitions between character states are all identical); SYM (symmetrical rates, in which forward and reverse transitions are equal); and ARD (all rates different, the maximum model, allowing different rates between all state transitions). More advanced users can also specify the rate matrix directly by, for example, specifying some transitions to be less likely than others.

We can explore the performance of the different models by extracting the scaled likelihoods at each node:

```
anc.ER<-ace(K, tree, type = "discrete", model="ER")$lik.anc
anc.SYM<-ace(K, tree, type = "discrete", model="SYM")$lik.anc
anc.ARD<-ace(K, tree, type = "discrete", model="ARD")$lik.anc
```

We can then plot the relative support for each state on the tree using pie charts at each node (fig. 5.7):

```
co <-c("blue", "yellow", "green")

plot(tree, show.tip.label=F)
nodelabels(pie = anc.ER, piecol = co, cex = 1)
title("ER reconstruction")

plot(tree, show.tip.label=F)
nodelabels(pie = anc.SYM, piecol = co, cex = 1)
title("SYM reconstruction")
```

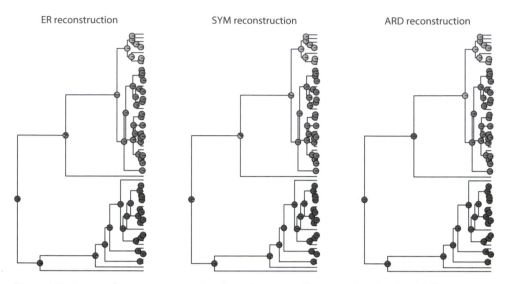

ER reconstruction   SYM reconstruction   ARD reconstruction

Figure 5.7. Ancestral state reconstruction for a three-state discrete trait using four different reconstruction methods (see text). Shading indicates trait state.

```
plot(tree, show.tip.label=F)
nodelabels(pie = anc.ARD, piecol = co, cex = 1)
title("ARD reconstruction")
```

Finally, we introduce the method of stochastic character mapping (Nielsen 2002), in which character histories are mapped on to the tree topology assuming a given substitution model under a continuous-time Markov process (Huelsenbeck et al. 2003). State frequencies can then be sampled to derive posterior probabilities for nodes. Bollback (2006) presents a simple heuristic describing the approach:

1. Calculate the conditional likelihood for each character state at each node of the tree.
2. Simulate ancestral states at each internal node by sampling from the posterior distribution of states.
3. Simulate the substitution history by sampling from the posterior distribution conditioned on the reconstructions in step 2 and the observed states at the tips of the topology.

Liam Revell has a user-friendly implementation in his *phytools* library, which follows the syntax in `ace()`, with an additional parameter `nsim` specifying the number of stochastic character maps:

```
simmap.trees<-make.simmap(tree, K, model="SYM", nsim=99)
```

Borrowing a function from Liam Revell,[2] we can estimate state frequencies at each node across the set of simulations:

2. Liam Revell maintains an excellent website (http://blog.phytools.org/) with many useful functions and updates not yet included in *phytools*.

SIMMAP reconstruction

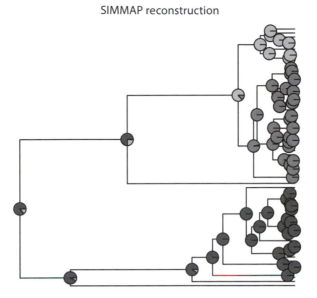

Figure 5.8. Ancestral state reconstruction for a three-state discrete trait using stochastic character state mapping (simmap()) (see text); shading indicates estimated state frequencies.

```
# function to compute the states
  foo<-function(x){
  y<-sapply(x$maps,function(x) names(x)[1])
  names(y)<-x$edge[,1]
  y<-y[as.character(length(x$tip)+1:x$Nnode)]
  return(y)
  }
```

We can then map these back on to the tree topology (fig. 5.8), as we did for our reconstructions using `ace`:

```
AA<-sapply(simmap.trees,foo)
piesA<-t(apply(AA,1,function(x,levels,Nsim) summary(fac-
      tor(x,levels))/Nsim,levels=c(0,1,2), Nsim=99))
plot(tree,show.tip.label=F)
nodelabels(pie=piesA,cex=1,piecol=co)
title("SIMMAP reconstruction")
```

One advantage to stochastic character mapping is that it can incorporate uncertainty in tree topology, branch lengths, and substitution models because we can calculate the joint posterior probability distribution of all parameters and then look at the marginal posterior probability distribution of specific parameters or summary statistics (Huelsenbeck et al. 2003).

5.4. CONCLUSION

A first step in any analysis should be to look at the data. When analyzing data within a phylogenetic framework, this should also include evaluation of its phylogenetic structure. We have reviewed common metrics for evaluating phylogenetic signal, and comparing fit of alternative evolutionary models. When trait data deviates significantly from assumptions of Brownian motion, the phylogenetic distances separating taxa on a time-calibrated tree might not accurately capture phenotypic distance between species. If we are able to identify the correct model of trait evolution, we can transform the branch lengths on the tree to match. Nonetheless, we might still favor using the untransformed tree because complex evolutionary models might not generalize across traits, and a BM model may be a better representation of aggregate trait differences when traits are evolving independently and idiosyncratically (but are phylogenetically structured). However, if the key trait structuring community coexistence is known *a priori*, we might wish to use the transformed tree; this would be the case if trait data are not available for all species in a community, or if we wish to describe community composition for an unmeasured function with a known phylogenetic structure. Finally, information on the evolutionary model and trait values for species within a community can help reveal evidence of competition and evolutionary character displacement.

Most phylogenetic methods assume the phylogeny is known (and can be represented by a single bifurcating tree) and species traits are measured without error. Both assumptions are rarely true. In addition, we are often beset by missing trait data for species and missing species from phylogenies. We have discussed how phylogenetic resolution can bias estimates of phylogenetic conservatism, and provided one solution. It is also now possible to incorporate measurement error in estimates of phylogenetic signal (Hardy and Pavoine 2012). Bayesian approaches allow us to sample over uncertainty in tree topology and evolutionary models, and perhaps provide the most promising way forward. We recommend that any researcher interested in addressing these questions consider carefully the alternative options available to them and, when possible, compare methods. Substantial differences in estimates between methods should be examined carefully.

⬧⬧

CHAPTER 6

The Geography of Speciation and Character Displacement

*The principle of divergence, plays, I believe, an important part
in the origin of species. The same spot will support more life
if occupied by diverse forms: we see this in the many
generic forms in a square yard of turf.*
—Charles Darwin, in Life and Letters (1887)

Gause's competitive exclusion principle predicts that no two species can coexist indefinitely on the same limiting resource; however, not just two, but hundreds of species coexist in natural communities. Competition models, such as the classic Lotka-Volterra equation, predict that the likelihood of competitive exclusion increases with the strength of interspecific competition (Harper 1978, Ricklefs 1979, May 1981). We might therefore expect coexisting species to occupy different ecological niches (as we have discussed in previous chapters), and thus demonstrate differences in their traits related to shared resource use; in turn, overdispersion in ecological traits has been interpreted as evidence for competition (Brown and Wilson 1956, Hutchinson 1959, Grant 1972). Evaluating the importance of competition in structuring ecological communities proved surprisingly problematic, and the sometimes acrimonious debate on the appropriate use and abuse of null models dominated ecology in the 1970s and 80s (Lewin 1983; see also chapter 4). Phylogenetic approaches come somewhat as a late entry into the field, but offer a novel, explicitly evolutionary perspective, and provide a simple alternative null model for exploring the relationship between coexistence and evolutionary divergence (Barraclough, Nee, et al. 1998, Davies et al. 2007).

Competition might structure the distribution of species traits via two general processes. First, selection might drive the divergence of ecological phenotypes by negative antagonistic interactions that reduce competition between sympatric populations utilizing the same resource base; here we refer to this as ecological character displacement (Brown and Wilson 1956, Taper and Case 1992). For the second process, ecological divergence occurs in allopatry, and competition filters species into communities based upon their trait values; here we refer to this as community-wide character displacement (see chapter 3). Traditionally, ecological character displacement has been assessed by contrasting trait values in sympatric versus allopatric populations, because we would predict larger differences between species when they occur sympatrically. Typically, community-wide character displacement is evaluated by comparing the dispersion in trait values among co-occurring species to a null distribution constructed by sampling species at random from the regional species pool (Strong et al. 1979, Gotelli and Graves 1996; see also chapters 3 and 4). Here we demonstrate how phylogenetic approaches can address both processes (ecological character displacement and community-wide character displacement), and in some cases help differentiate between them. This chapter builds on the material covered in chapter 3 by examining broad-scale co-occurrence patterns.

6.1. CHARACTER DIVERGENCE AND GEOGRAPHIC OVERLAP

The geographic distribution of a species reflects multiple factors, including its strength of phylogenetic niche conservatism (chapter 5) and the geography of present day climate, its dispersal ability, and the history of speciation. In addition, biotic factors, including predators, symbionts, and competitors, might also limit a species' geographical distribution. One explanation for why invasive species are able to spread so rapidly is because humans have assisted them in overcoming biogeographical barriers (either deliberately or by accident) and they have thus been able to escape their native predators and competitors (Blossey and Notzold 1995). In this chapter we are interested in evaluating the effect of competition in the native range. We have seen how phylogenetic community structure can help provide insights to the processes configuring ecological communities, and how our interpretation of phylogenetic patterns is predicated on the strength of evolutionary conservatism among relevant characters (chapter 5). Implicit within such models is an assumption that the evolution of a particular trait in one lineage is more or less independent of the species values for the same trait in co-occurring species. Here we consider a scenario in which the evolution of species traits may itself be a product of species interactions.

Let us consider the simplest case. This would be two sister species that overlap to some degree in their geographic distribution. Because sister species are each other's closest relatives, and thus likely to share similar resource requirements and life histories, we might predict a high potential for competition between them. For our purposes here, we consider the extent of range sympatry as a measure of potential for interspecific contact, and thus possible competition (following Diamond 1986, Barraclough, Nee, et al. 1998), although habitat occupancy and population density would obviously be important also. We can derive three broad hypotheses relating to contemporary geographical overlap between species pairs (see also Barraclough, Nee, et al. 1998), and differentiate among them by examining the predicted relationships with sympatry, phylogenetic relatedness, and ecological disparity.

Hypothesis (1) is the null expectation which assumes that (a) shifts in species ranges since divergence have been extensive, (b) competitive interaction between contemporary species pairs is weak or absent, and (c) a species location in space is independent of the abiotic environment. Contemporary overlap between two sister species would therefore reveal no information on species interactions or mode of speciation. Species range overlap would be solely a function of the geometric constraints imposed by the respective range sizes and the relative size of the domain in which they occur, as predicted by, for example, mid-domain models of species richness (Colwell and Hurt 1994).

Hypothesis (2), the evolutionary explanation, assumes the divergence of ecological phenotypes driven by mutually negative interaction as a response to selection acting to reduce competition; the result would be evolutionary character displacement (Brown and Wilson 1956, Taper and Case 1992). If we consider two populations of a single species (rather than two competing species), it is possible that competition for resources might be sufficient to drive speciation in sympatry (Diekmann and Doebeli 1999, Day and Young 2004), with divergent selection driving rates of character change (Schluter 2000, Coyne and Orr 2004).

Sympatric speciation remains a controversial topic and biogeographical evidence supports the predominance of a geographical (allopatric) mode of speciation (Mayr 1963, Lynch 1989). However, recent theoretical and empirical work has generated renewed interest in the subject (Turelli et al. 2001, Via 2001, Bolnick 2004, Coyne and Orr 2004). Alternatively, speciation may be allopatric; range shifts result in secondary sympatry and subsequent character displacement. The two cases may be represented as follows:

a. Sympatry (incipient species) → ecological divergence → speciation (sympatric)
b. Speciation (allopatric) → sympatry → ecological divergence

Time flows from left to right; the arrows indicate direction of causality.

Hypothesis (3), the ecological explanation, assumes speciation and ecological divergence occurs in allopatry; sympatry is the product of subsequent independent range shifts or dispersal over time. This case may be represented as follows:

c. Speciation (allopatric) → ecological divergence → sympatry

This last hypothesis is the closest match to assumptions underlying most analyses of phylogenetic community structure (e.g., chapter 3); but see Cadotte and Jin (2014).

The mode of speciation can differentiate between case (a) and case (b) or (c). Both case (b) and case (c) predict greater sympatry with increasing ecological divergence, although the direction of causation differs. However, for character divergence in allopatry, as in case (c), we might predict that rates of trait change would be similar between sympatric and allopatric species pairs. In contrast, if selection in sympatry drives divergence, as in cases (a) and (b), we would predict that two sympatric sister species would demonstrate greater trait differences than two allopatric sister species.[1]

6.1.1. Terrestrial Carnivores

Testing the hypotheses that relate geographical overlap to species divergence is relatively straightforward, and requires information only on species range overlap, trait values and, of course, phylogeny. Here we reanalyze the data on terrestrial carnivore range overlap and dentition from Davies et al. (2007). First, as usual we read in the phylogeny using the *ape* library, and then use functions from the *geiger* library to obtain divergence times between sister pairs:

```
library(ape)
library(geiger)
tree<-read.tree("carnivores_chrono.phy")
```

From this tree we can identify sister species pairs by extracting the list of species descending from each internal node, using the `tips()` function from *geiger*:

```
tips(tree,x)
```

Here x is the specified node. It is useful here to remember how *ape* indexes nodes on objects of class `phylo` (i.e., phylogenetic trees), numbering starts at the tips (i.e., species), which are labeled from 1 to *n* total number of tips on the tree; internal node numbers therefore

---

1. Jonathan thanks Patrick Nosil for pointing out this expectation to him in a review he provided on one of Jonathan's publications; this is an illustration of how the review process can help advance our thinking.

start at $n + 1$ (see chapter 2 for further details). In our example of carnivore phylogeny there are 268 tips:

```
length(tree$tip.label)
```

And there are 209 internal nodes:

```
Nnode(tree)
```

On a fully bifurcating, rooted tree there are $n - 1$ internal nodes, and we are therefore "missing" 59 nodes; these are evident as polytomies (i.e., unresolved nodes) in the phylogeny, which can be easily visualized by plotting the tree. By definition, nodes subtending sister species have two descendants. We can therefore run through each internal node in the tree to extract the list of descendant species, and then exclude all nodes with >2 descendants to return the list of sister pairs, here stored in an object named *sister.pairs*. In addition, we are interested in their divergence times, because we wish to evaluate whether sympatric sister taxa have diverged more rapidly than their allopatric counterparts. Divergence times can be obtained using the node.depth.edgelength() function in *ape*; however, this returns the depth of the node from the root. To return node depth from the present, we can simply subtract node.depth.edgelength from the total age (depth) of the tree:

```
tree.age<-max(cophenetic(tree))/2
node.ages<-tree.age-node.depth.edgelength(tree)
```

The cophenetic() function returns all the pairwise distances between tips, and underlies many phylogenetic dispersion functions (chapter 3). Here we extract the maximum pairwise distance, as this must go through the root node, and then we divide by 2 because we wish to return only the distance from tip to root. (Note that this shortcut is only appropriate for ultrametric trees.)

We can now write a simple "for" loop to run through the internal nodes on the tree (starting from node = n), identify sister pairs, and record their divergence times:

```
nodes<-Nnode(tree)
species<-length(tree$tip.label)

sister.pairs<-NULL
age<-NULL

i = 1
for (x in species:(species+nodes)) {
sis<-tips(tree,x)
if(length(sis)==2){
sister.pairs<-rbind(sister.pairs, sis[1:2])
age[i]<-node.ages[x]
i = i +1
}#end if
}#end for
```

We then combine this information into a data frame:

```
data<-data.frame(age, I(sister.pairs[,1]), I(sister.
     pairs[,2]))
colnames(data)<-c("age", "sisterA", "sisterB")
```

Note here the use of I() before the sister.pairs[]; this keeps the data class in its original format (character string). Without specifying this command, species names would have been converted into class type = factor. In this instance, the difference between a factor and a character string is not important, but it is good practice not to confuse data classes, especially when handling class numeric. As an illustration, try the following:

```
x<-4
x<-as.factor(x)
x<-as.numeric(x)
x
```

Next we need information on the relative overlap between sister taxa. This data was generated by overlaying species distribution maps from Grenyer et al. (2006). There are a number of libraries in R that will handle GIS data (e.g., http://cran.r-project.org/web/views/Spatial.html); however, we will skip quickly over this step, and use the published values that are available from http://press.princeton.edu/titles/10775.html.

```
geog.data<-read.table("carnivore range data.txt", header=T)
```

Merging the geographic data with the sister species pairs is a little more complicated than usual because we are matching data to pairs of sister taxa. To simplify this step, we can identify the species in each sister pair that comes first alphabetically. Here we have used the min() function (for data of type character, it sorts values lexicographically), and applied this across the rows of both data sets:

```
phylo.sis.pair<-apply(data[,c("sisterA", "sisterB")], 1, min)
geo.sis.pair<-apply(geog.data[,c("Species_A", "Species_B")],
        1, min)
```

We now have species we can match between the two data sets to merge our sister pairs with the data on range overlap. We first reorder the data frame we created with sister pairs and their divergence times using the match() function, and then simply bind the two data sets together using cbind():

```
carnivores<-cbind(geog.data, data[match(geo.sis.pair, phylo.
        sis.pair),])
```

Range size and range overlap are in units of millions km², and we therefore want to convert overlap into an index of proportional sympatry; following Barraclough, Vogler, et al. (1998) we divide the area of overlap by the range size of the species with the larger range:

```
max.range<-apply(carnivores[,c("Range_size_A", "Range_
        size_B")], 1, max)
sympatry<-carnivores$Overlap/max.range
carnivores<-cbind(carnivores, sympatry)
```

We can now examine the relationship between divergence time and sympatry in our sister species pairs (fig. 6.1). Because our index of sympatry is bounded between one and zero, we arcsine transform it prior to fitting our linear regression model. We could alternatively have used a generalized linear model (GLM) with a binary error structure, but we will stick with linear regression to match the original analyses presented in Davies et al. (2007):

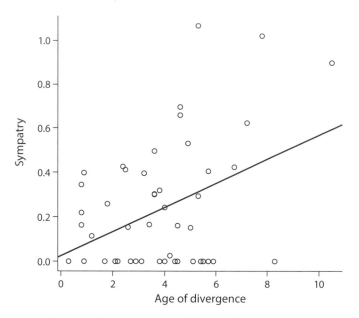

Figure 6.1. Regression between divergence time and degree of range sympatry among sister species of carnivores.

```
model<-lm(asin(sqrt(sympatry))~age, data = carnivores)
summary(model)
Call:
lm(formula = asin(sqrt(sympatry)) ~ age, data = carnivores)

Residuals:
     Min       1Q    Median       3Q       Max
-0.47657  -0.18284  -0.03029  0.18584   0.75643

Coefficients:
             Estimate  Std. Error  t value  Pr(>|t|)
(Intercept)   0.02509     0.07699    0.326   0.74593
age           0.05440     0.01763    3.086   0.00337 **
---
Signif. codes: 0 '***' 0.001 '**' 0.01 '*' 0.05 '.' 0.1 ' ' 1

Residual standard error: 0.2611 on 48 degrees of freedom
Multiple R-squared: 0.1655, Adjusted R-squared: 0.1481
F-statistic: 9.521 on 1 and 48 DF, p-value: 0.003367

plot(carnivores$age, asin(sqrt(sympatry)))
abline(model)
```

We find a highly significant relationship between divergence time and sympatry, with greater geographical overlap among sister pairs that diverged longer ago. The relationship shows large scatter, and certainly the amount of variation in overlap explained by divergence time is low, at around 17%. Nonetheless, this simple plot contains a lot of important information. First, there are no recently diverged species showing large geographical overlap; this suggests, perhaps unsurprisingly, that the geographical mode of speciation was likely predominantly allopatric. The interpretation of these age-range correlations has, however, been a matter of some debate. For example, Chesser and Zink (1994) and Losos and Glor (2003) argue that postspeciational range shifts would effectively mask any relationship between contemporary distributions and the geographic mode of speciation. Undoubtedly, many species will have large shifts in their distributions, most obviously during glacial-interglacial cycles. We should thus regard such correlations with healthy cynicism, and certainly the suggestion that any evidence of contemporary range overlap might provide evidence of sympatric speciation (e.g., Mattern and McLennan 2000) would seem rather overenthusiastic. Nevertheless, it is unclear why we might expect a significant relationship between time of divergence and range overlap if speciation was predominantly sympatric.

It is possible to construct null models of range shifts over time, much as we might model the evolution of a morphological character in trait space (chapter 5); but instead we could model the geographical drift of range location and various modes of speciation, as described in Barraclough and Vogler (2000).[2] We would then be able to compare our observed age-range correlations to these null expectations. The *phyloclim* library provides an alternative significance test of age-range correlations (`age.range.correlation()`) following Fitzpatrick and Turelli (2006), in which observed overlap between sister clades is compared to expectations derived from Monte Carlo simulations permuting the matrix of age and average overlap for each node. Because ranges can be labile, and there are inherent difficulties in estimating ancestral distributions (Losos and Glor 2003), comparisons across sister clades deeper in the tree will tend to be weaker; if there are sufficient data it may be preferable to focus on sister species and/or more recent divergence events (Avise 2000).

From our plot of the data (fig. 6.2) we can also observe a number of species pairs that have remained in allopatry despite apparently long times since divergence. Once again, this pattern might reflect the general dominance of allopatric speciation, perhaps via vicariance biogeography in which a biogeographic boundary not only provides reproductive isolation during speciation, but subsequently maintains geographic isolation between species; then, range overlap is not possible, irrespective of species biotic interaction strengths. If we exclude these noninteracting species pairs, we can explore in more detail the relationship between divergence time and potential for sympatry:

```
model<-lm(asin(sqrt(sympatry))~age, data = subset(carnivores,
    sympatry>0))
summary(model)
Call:
lm(formula = asin(sqrt(sympatry)) ~ age, data = subset
    (carnivores,
sympatry > 0))

Residuals:
    Min        1Q    Median       3Q       Max
-0.39345  -0.15185  -0.00746  0.12071  0.57329
```

2. Albert Phillimore at the University of Edinburgh wrote R code to do just this, and promises to provide it on request.

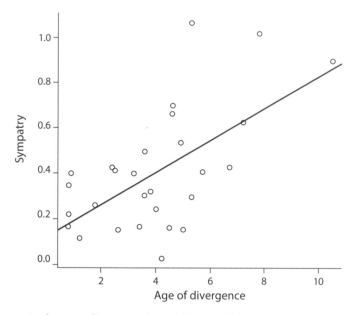

Figure 6.2. Regression between divergence time and degree of range sympatry among sister species of carnivores as in figure 6.1, but excluding completely allopatric species pairs.

```
Coefficients:
              Estimate  Std. Error   t value    Pr(>|t|)
(Intercept)    0.12075     0.08059     1.498    0.145673
age            0.07090     0.01767     4.012    0.000429 ***
---
Signif. codes:  0 '***' 0.001 '**' 0.01 '*' 0.05 '.' 0.1 ' ' 1

Residual standard error: 0.2125 on 27 degrees of freedom
Multiple R-squared: 0.3734, Adjusted R-squared: 0.3502
F-statistic: 16.09 on 1 and 27 DF, p-value: 0.0004293

plot(asin(sqrt(sympatry))~age, data = subset(carnivores,
        sympatry>0), ylab = "Sympatry", xlab = "Age of
        divergence")
abline(model)
```

In this new plot we can clearly see a much stronger relationship between divergence time and range overlap, suggesting that over time, species pairs have increasingly come into secondary sympatry, and overlap more in their geographic ranges. We now explain a much more impressive 37% of the variation in sympatry between sister taxa. We thus gain additional insight into the evolutionary and biotic processes limiting species range distributions. First, an allopatric mode of speciation is even more strongly supported. Second, absence of range overlap among some sister pairs, even after long periods of time since divergence, suggests the presence of geographical barriers separating species; this is perhaps indicative of vicariance speciation. Third, when species have come into secondary contact, range overlap is initially low, and extensive overlap is only observed among more distantly related sister pairs. One question we might therefore wish to ask is whether the correlation between age and overlap simply reflects independent range dynamics, in which species pairs wander more or less by chance into secondary sympatry following speciation in allopatry, or whether competition inhibits large overlap between more closely related taxa. Both processes might generate patterns of phylogenetic overdispersion; however, the former would better reflect the available species pool from which the community samples (see chapter 3), whereas the latter is a product of competitive exclusion among close relatives, and fits more closely with traditional interpretations of phylogenetic overdispersion (Webb et al. 2002). Regardless, this general pattern has important implications for the community level analyses presented in chapter 3, and raises some interesting questions. If regional species pools generally contain species capable of coexisting at some scale but have not yet come into sympatry, what does this imply when we search for phylogenetic patterns? The species pools used in null models (chapter 4) may be biased by the geography of speciation. Further, as non-native species are moved across biogeographical barriers, how do we expect relatedness to influence species interactions?

For carnivores, evidence of competition is strong, and in some cases may take the form of direct interference or intraguild predation (Polis et al. 1989). Further, a number of studies have shown support for niche partitioning within carnivore guilds (e.g., Simms 1979; Dayan et al. 1989, 1990; Dayan and Simberloff 1994; Van Valkenburgh and Wayne 1994; Palomares and Caro 1999; Van Valkenburgh 1999; Loveridge and Macdonald 2002, 2003), indicating that competition might inhibit co-occurrence among species occupying similar niches. If competition limits secondary sympatry, we would then predict differences in resource use to be the primary determinant of range overlap. In contrast, if secondary sympatry is a product of range dynamics independent from biotic interactions, we might predict time of divergence to be the better predictor of range overlap. In carnivores, dental morphology is closely linked with diet; while canines are perhaps the most distinctive teeth, they have been subject to strong sexual selection, as evident in their sometimes considerable sexual dimorphism. We therefore focus here on the carnassials, which can be either bladelike for slicing meat, or blunt cusped for cracking bones or masticating plant matter. For species that are predominantly carnivorous, carnassial size is known to correlate with prey size (Dayan et al. 1989, 1992; Dayan and Simberloff 1998), and can thus provide an indicator for niche partitioning.

We can explore the relationship between range overlap and trait divergence in much the same way that we evaluated divergence time. The first step is to match the trait data to the species pairs, here again using data from table S1 in Davies et al. (2007), and following the same steps as above:

```
teeth.data<-read.table("carnivore teeth data.txt", header=T)
teeth.sis.pair<-apply(teeth.data[,c("Species_A", "Spe-
      cies_B")], 1, min)
carnivores.traits<-cbind(carnivores, teeth.data[match
      (geo.sis.pair,teeth.sis.pair),])
```

Note that in this case the order of sister pairs matches exactly that in our carnivore data frame, so the match() function was unnecessary; nonetheless, it is always good practice not to assume separate data tables are ordered similarly. The data set contains information on the lower carnassial (M1), length of the upper carnassials (PM4), and canine diameter.

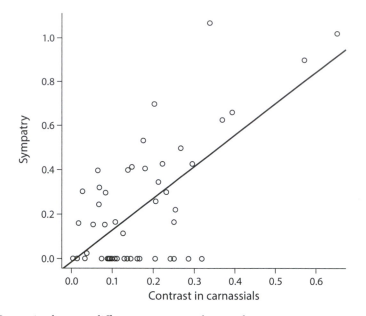

Figure 6.3. Regression between difference in carnassial size and range sympatry among sister species of carnivores (compare with figure 6.1).

We will focus on the upper carnassials, but you can explore relationships using the other dental measures. If competition is restricting geographic overlap between sister pairs, we would predict less range overlap among sister species with similar dentition, so as a first step we need to calculate the absolute difference in trait values between the sister species, here using the logarithm of PM4 length:

```
carnassial<-abs(log(carnivores.traits$PM4_A)-log(carnivores.
    traits$PM4_B))
carnivores.traits<-cbind(carnivores.traits, carnassial)

model<-lm(asin(sqrt(sympatry))~carnassial, data = carni-
    vores.traits)
```

Again, we can generate a simple plot of the data and fit the regression line from our linear model (fig. 6.3):

```
plot(asin(sqrt(sympatry))~carnassial, data = carnivores.
    traits, ylab = "Sympatry", xlab = "Contrast in
    carnassials")
abline(model)
```

Now let us look at the model fit:

```
> summary(model)

Call:
lm(formula = asin(sqrt(sympatry)) ~ carnassial, data =
    carnivores)
```

```
Residuals:
     Min        1Q     Median        3Q        Max
-0.43942   -0.13506   -0.00982   0.12763    0.60194

Coefficients:
              Estimate  Std. Error   t value  Pr(>|t|)
(Intercept)  -0.01591     0.05020     -0.317     0.753
carnassial    1.42873     0.23013      6.208   1.2e-07  ***
---
Signif. codes:  0 '***' 0.001 '**' 0.01 '*' 0.05 '.' 0.1 ' ' 1

Residual standard error: 0.2129 on 48 degrees of freedom
Multiple R-squared: 0.4454,  Adjusted R-squared: 0.4338
F-statistic: 38.54 on 1 and 48 DF,  p-value: 1.204e-07
```

We can now explain over 44% of the variation in range overlap, and this increases to almost 63% if we exclude completely allopatric sister species pairs on the assumption that geographical barriers might inhibit sister species from coming into secondary contact (fig. 6.4):

```
model<-lm(asin(sqrt(sympatry))~carnassial, data = subset(-
    carnivores.traits, sympatry>0))

plot(asin(sqrt(sympatry))~carnassial, data = subset(carni-
    vores.traits, sympatry>0), ylab = "Sympatry", xlab =
    "Contrast in carnassials")
abline(model)
```

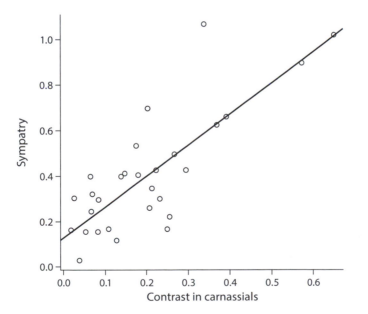

Figure 6.4. Regression between difference in carnassial size and sympatry among sister species of carnivores, excluding completely allopatric species pairs (compare with figure 6.3).

The model output can be displayed as before using the `summary()` function:

```
summary(model)

Call:
lm(formula = asin(sqrt(sympatry)) ~ carnassial, data = sub-
    set(carnivores.traits,sympatry > 0))

Residuals:
     Min        1Q    Median        3Q       Max
-0.30366  -0.10463  -0.00310   0.08092   0.48129

Coefficients:
             Estimate  Std. Error  t value  Pr(>|t|)
(Intercept)   0.12662     0.05081    2.492    0.0191  *
carnassial    1.36411     0.20141    6.773  2.84e-07  ***
---
Signif. codes: 0 '***' 0.001 '**' 0.01 '*' 0.05 '.' 0.1 ' ' 1

Residual standard error: 0.1634 on 27 degrees of freedom
Multiple R-squared: 0.6295, Adjusted R-squared: 0.6158
F-statistic: 45.87 on 1 and 27 DF, p-value: 2.839e-07
```

From the scatter plot (fig. 6.4), we can see one clear outlier with much greater geographical range overlap than predicted from differences in carnassials. We can query which record this is using the `identify()` function, and clicking on the outlier in the R-plot window:

```
identify(asin(sqrt(sympatry))~carnassial, data = subset
    (carnivores.traits, sympatry>0))
```

This point is shown to be for record 12, the contrast between the ocelot (*Leopardus pardalis*) and margay (*Leopardus wiedii*). Both species are small, nocturnal felines in South America, and it is not immediately clear how they are able to persist in sympatry. It has been suggested that low numbers of ocelots allows a higher density of margays, indicative of interspecific competition (De Oliveira et al. 2010). However, both species occur at relatively low densities, and will take a wide range of prey items, including small mammals, birds and their eggs, fish, amphibians, reptiles, arthropods, and even fruit in the case of the margay (Nowak 1999). It is possible, therefore, that direct competition for resources is relatively rare. In addition, it is thought that the lighter margay forages for food in trees (Konecny 1989), whereas the larger and heavier ocelot is mostly terrestrial, further reducing potential for competition.

We have shown that range overlap among sister species of carnivores increases with time since divergence, consistent with a model of allopatric speciation followed by secondary sympatry in some species. Whether two species come into secondary sympatry is likely first determined by geographical barriers to range distributions, while the extent of range overlap is proportional to differences in functional traits related to resource use (here, prey size); this fits with a model of competitive exclusion among species occupying similar niches. However, as we discussed above, there are two alternative mechanisms that give rise to the positive relationship between range overlap and trait divergence. The first is evolutionary character displacement driven by selection to reduce competition between sympatric species. The second is ecological species sorting, in which species diverge in allopatry, and competition is only important later when species come into secondary contact, thus inhibiting range overlap among species occupying similar ecological niches.

One insight can be gained by looking back to the plot of contrasts in carnassials against range overlap, including allopatric species pairs (fig. 6.3). When we examine the difference in carnassial sizes between sympatric and allopatric species pairs we can see that contrasts in carnassials are never greater than 0.35 for allopatric species pairs; whereas among sympatric species pairs with over 50% overlap in their distributions, contrasts in carnassials are much greater, ranging from 0.5–6.5. If species sorting predominates, we would predict that large differences in traits between allopatric sisters would be just as likely as small differences. In contrast, if evolutionary character displacement drives species differences, we might predict faster rates of divergence among sympatric species due to the additional effect of competition on trait divergence (Schluter 2000). Our results, showing large differences only in sympatric pairs, would support a model of evolutionary character displacement.

6.2. COMMUNITY-WIDE TRAIT DISPERSION

Our models of phylogenetic community structure discussed in chapter 3 are typically used to infer the process of community assembly, such as filtering and competition. Here, we have discussed the possibility that community composition might shape the distribution of traits via divergent selection (evolutionary character displacement). Traditionally, evidence for evolutionary character displacement has been assessed by contrasting differences between sympatric and allopatric populations. In contrast, community-wide character displacement as indexed by, for example, the variance in size ratios (Holmes and Pitelka 1968, and see below) or similar dispersion statistics, has been evaluated by contrasting the observed distribution of traits against a null generated for sampling some predefined species pool (see also chapter 4). However, as we have shown for metrics of *NTI* and *NRI* (chapter 3), our inferences can be highly sensitive to how we delineate the species pool and how we sample from it (chapter 4). Phylogenetic methods allow us to explore evidence for evolutionary character displacement within multispecies communities, and offer an alternative approach for generating a null distribution from which to contrast empirical observations; these methods are based on evolutionary expectations derived from phylogenetic relationships among community members.

Here, staying with mammals, we use a community data set from Yotvata, Israel, including a phylogenetic tree (fig. 6.5):

```
yotvata<-read.tree("yotvata.tre")
plot(yotvata)
```

Data on body sizes is also included:

```
traits<-read.table("yotvata data.txt",header=T)
```

We can generate a variety of trait dispersion indices directly from the observed trait data. Perhaps one of the simplest community-wide metrics is the variance in size ratios (VSR), mentioned above, which reflects the evenness in spacing between the natural logarithm of ranked trait values for species within a community (Holmes and Pitelka 1968).

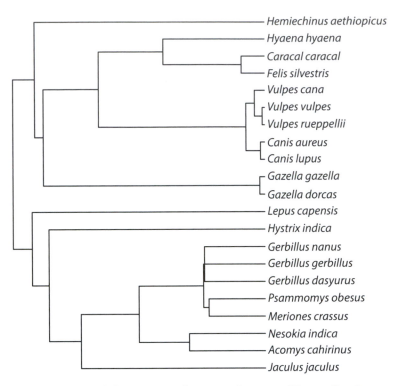

Figure 6.5. Phylogenetic tree for mammal species of Yotvata, Israel.

First, we simply sort the data, here mass measured in grams, and then log-transform the values:

```
mass<-log(sort(traits$mass))
```

Next we can calculate the variance in size ratios by using a short "for" loop to calculate the difference in logged mass between each pair of adjacent values:

```
myvar<-numeric()
for (i in length(mass):2) {
myvar[i-1]<-mass[i]-mass[i-1]
}# end loop
```

Then we simply calculate the variance of the differences:

```
VSR<-var(myvar)
```

Here we have a variance in size ratios of 0.073. If competition is high and traits are reasonably labile we would predict that species would tend to show more even spacing in their trait values related to resource utilization to minimize niche overlap. Thus, low VSR, which reflects constant size ratios, is thought to reflect interspecific competition (Hutchinson 1959, Grant 1972, Simberloff and Boecklen 1981), and can be interpreted in much the same way as we have traditionally interpreted patterns of phylogenetic overdispersion (chapter 3). In contrast, large VSR is indicative of uneven trait spacing or clustering. Metrics exploring community-wide phylogenetic clustering, such as MPD (chapter 3), are powerful for detecting clustering within a single clade, for example, as might be expected if all species within a community were filtered on a single, strongly phylogenetically structured trait. However, VSR is more sensitive for detecting multimodality in size distributions, with species clustering around distinct "optima." Such discontinuities in trait distributions may be common within species assemblages (Holling 1992), and might reflect filtering into discrete niches or multiple peaks on the fitness landscape. Where does our observed value of 0.073 fall on this scale? We need to compare our value against some null distribution to assess whether traits values are more or less aggregated than expected by chance.

Traditionally, the null expectation has been generated by randomly sampling from the regional pool of potential community members. There is an extensive literature on the selection of appropriate null models in ecology, reviewed in detail by Gotelli and Graves (1996) and discussed in chapter 4. As a first pass, we can generate our null distribution for carnivores by assuming a pool of all species in Israel, and generate null VSR by randomly sampling species from the country pool and recalculating VSR, holding the number of species constant. We repeat this procedure multiple times to generate our expected distribution:

```
Israel<-read.table("Israel.mammals.txt", header=T)

sample.VSR<-numeric()
rand.var<-numeric()

for (x in 1:999){
sample.mass<-log(sort(sample(Israel$mass, length(mass))))

for (i in length(sample.mass):2) {
rand.var[i-1]<-sample.mass[i]-sample.mass[i-1]
}# end i loop
sample.VSR[x]<-var(rand.var)

}# end x loop
```

We can then compare whether our observed VSR falls in the tails of the null distribution, following convention for a two-tailed test; we might consider our observed VSR as significant if more than 97.5% of randomized VSR values are higher, or if less than 2.5% of randomized VSR are lower. For illustration, we can plot our observed values on a histogram showing the distribution of randomized values:

```
hist(sample.VSR)
abline(v=VSR, col="red", lwd = 3)
```

Now let's plot where the tails of the distribution lie:

```
abline(v=quantile(sample.VSR, 0.025), lwd = 2, lty=2)
abline(v=quantile(sample.VSR, 0.975), lwd = 2, lty=2,)
```

We can see clearly (fig. 6.6) that our empirical estimate of VSR falls outside the 95% limits of our expectations derived from resampling the data at random, and we might therefore conclude that the mammal community in Yotvata is structured by competition (VSR is significantly lower than our null expectation) and leave it there. However, as we have seen in chapter 4, conclusions can change dramatically with the extent of the species pool from which we draw; communities that appear overdispersed at finer scales

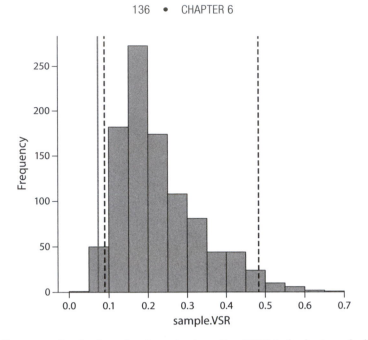

Figure 6.6. Frequency distribution of variance in size ratios (VSR) in body size calculated from a random draw of Israeli mammals in comparison to the empirical value from the observed mammal community at Yotvata (*vertical line*). *Dashed lines* represent 95% confidence intervals.

can appear underdispersed at broader scales. If we are interested in detecting evidence of competition, we might want to focus on finer spatial and taxonomic scales (see, e.g., Cavender-Bares et al. 2009, Vamosi et al. 2009), but we still need a sufficient sample size of species in the pool to generate a statistically appropriate null distribution. If we draw from a very small pool, we will likely resample the same set of species multiple times, and our null distribution cannot be considered to be composed of truly independent draws. It is not clear how to determine the most appropriate scale for delineating the species pool. One approach is to explore patterns assuming different species pools. In an analysis of island mammal communities, Cardillo et al. (2008) explored phylogenetic structure assuming regional source pools at different scales. In this example, it was found that many islands demonstrated phylogenetic overdispersion, and the definition of the regional pool had little influence on patterns. However, we might consider island-mainland systems to already represent relatively large-scale processes (islands might encompass many thousands of km²) in comparison to coexistence models at the plot scale, which may be on the order of 1 m² or less.

Phylogeny provides us with an alternative null model that is independent of the regional species pool. We have already reviewed some common models of character evolution, including Brownian motion, where traits diverge over time independently in a manner analogous to a random walk (see chapter 5). Brownian motion therefore represents a neutral model of evolution, and we can thus use the known phylogeny linking species in a community to generate an expected distribution of traits for the species set. Previously we used the `rTraitCont()` function from *ape*; however, due to the ubiquity of the Brownian motion model in the phylogenetic literature there are many equivalent functions that we can use. Here we use the `fastBM()` function from Liam Revell's *phytools* library as it allows us to rapidly generate many traits simultaneously:

```
library(phytools)
sim.traits<-fastBM(yotvata,sig2=0.1, nsim=99)
```

And here `sig2` is the Brownian rate parameter, or more formally, the instantaneous variance of the BM process, and `nsim` specifies the number of independent simulated traits. We now have to calculate our VSR for each of the simulated traits, as we did for the regional species pool:

```
sim.var<-numeric()
sim.VSR<-numeric()
for (x in 1:99){
sim.mass<-sort(sim.traits[,x])

for (i in length(sim.mass):2) {
sim.var[i-1]<-sim.mass[i]- sim.mass[i-1]
}# end i loop
sim.VSR[x]<-var(sim.var)

}# end x loop
```

And again we plot the data:

```
hist(sim.VSR)
abline(v=VSR, col="red", lwd = 3)
abline(v=quantile(sim.VSR, 0.025), lwd = 2, lty=2)
abline(v=quantile(sim.VSR, 0.975), lwd = 2, lty=2,)
```

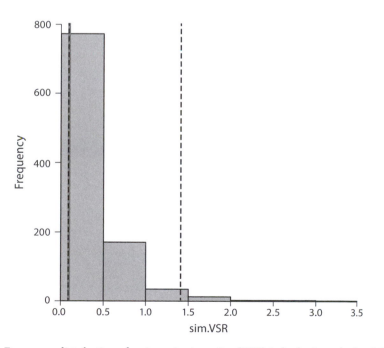

Figure 6.7. Frequency distribution of variance in size ratios (VSR) in body size calculated from simulating a Brownian motion trait on the known phylogeny for the mammal community at Yotvata, in comparison to the empirical VSR from measured data (*vertical line*). *Dashed lines* represent 95% confidence intervals. Brownian rate parameter for simulations (`sig2`) = 0.1.

From figure 6.7 we can see that the observed VSR falls to the left of the distribution from simulations, and just inside the 2.5% required for statistical significance; however, results are somewhat stochastic, as each simulation run will differ slightly. Note that we do not log the trait values before calculating differences in the simulated data; this is because we have simulated the traits under a model of pure drift in which traits evolve incrementally, and the change in trait value is independent of the size of the trait. In contrast, quantitative traits may be more likely to evolve under a log-Brownian process, where the raw change in trait value per unit time is likely to be larger for species with larger trait values. For example, if we consider body size in mammals, evolutionary change might be measured in grams for mice and kilograms for elephants.

However, we can see the distribution of VSRs from simulated data is highly right skewed, with a long tail representing a few very large VSRs. In addition, we can nudge the simulations such that our observed values appear more significant. We can do this by, for example, increasing the Brownian rate parameter discussed above (fig. 6.8):

```
sim.traits<-fastBM(yotvata,sig2=0.5, nsim=1000) # Brownian
        motion
```

In general, simulations with small `sig2` will return small VSR, whereas simulations with large `sig2` will return correspondingly large VSR. To overcome this bias, we can rescale the simulated data to the bounds of the empirical trait values using the `rescale()` function in the *plotrix* library (fig. 6.9):

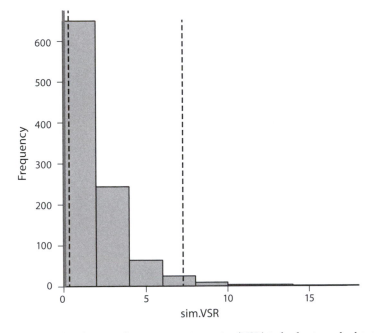

Figure 6.8. Frequency distribution of variance in size ratios (VSR) in body size calculated from simulating a Brownian motion trait on the known phylogeny for the mammal community at Yotvata, in comparison to the empirical VSR from measured data (*vertical line*). *Dashed lines* represent 95% confidence intervals. Brownian rate parameter for simulations (`sig2`) = 0.5. Compare with figure 6.7.

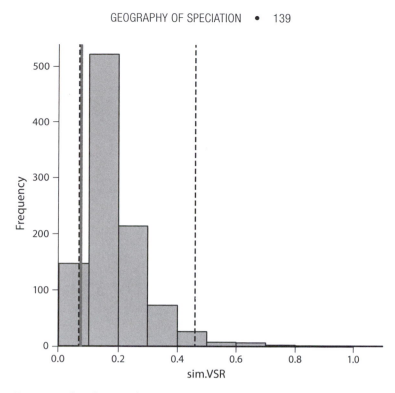

Figure 6.9. Frequency distribution of variance in size ratios (VSR) in body size calculated from simulated Brownian motion traits rescaled to the observed range of trait values for the mammal community of Yotvata. Empirical VSR from measured data is represented by the *vertical line. Dashed lines* represent 95% confidence intervals.

```
library(plotrix)#rescale function
sim.mass<-rescale(sort(sim.traits[,x]),range(VSR.dat))
```

Alternatively, we can rescale both empirical traits values to a mean of zero and a standard deviation of one, using the `scale()` function:

```
VSR.scale.dat<-scale(log(sort(mass)))
myvar<-numeric()
for (tree.tips in (length(yotvata$tip.label)):2) {
myvar[tree.tips-1]<-VSR.scale.dat[tree.tips]-VSR.scale.
        dat[tree.tips-1]
}# end simulation tree.tips loop
VSR.scale<-var(myvar)

sim.traits<-fastBM(yotvata,sig2=0.1, nsim=1000) # Brownian
        motion

sim.var<-numeric()
sim.VSR<-numeric()
for (x in 1:1000){
sim.mass<-scale(sort(sim.traits[,x]))
```

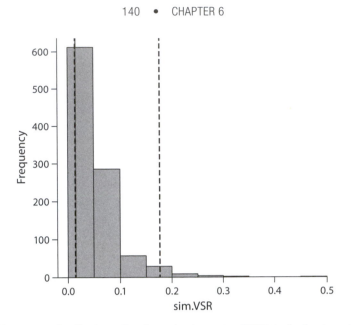

Figure 6.10. Frequency distribution of variance in size ratios (VSR) in body size calculated from simulated Brownian motion traits rescaled to a mean of zero and a standard deviation of one. Empirical VSR from rescaled measured data is represented by the *vertical line. Dashed lines* represent 95% confidence intervals.

```
for (i in length(sim.mass):2) {
sim.var[i-1]<-sim.mass[i]- sim.mass[i-1]
}# end i loop
sim.VSR[x]<-var(sim.var)

}# end x loop

hist(sim.VSR)
abline(v=VSR.scale, col="red", lwd = 3)
abline(v=quantile(sim.VSR, 0.025), lwd = 2, lty=2)
abline(v=quantile(sim.VSR, 0.975), lwd = 2, lty=2,)
```

In both the rescaled cases above we can see that the empirical VSR falls on or close to the 2.5% tail (figs. 6.9 and 6.10). This indicates that the distribution of body sizes in the Yotvata mammal community are more evenly spaced than predicted from a null model in which traits evolve along the branches of phylogeny according to a Brownian motion model. Reassuringly, our results match previous studies that have detected an imprint of competition on the structure of mammal communities in the region (e.g., Dayan et al. 1989, Dayan et al. 1990, Yom-Tov 1991).

There are two general classes of explanations for our findings. First, traits might actually evolve following a Brownian motion model, but the combination of species that can coexist represents the subset of species that are overdispersed. Remember that Brownian motion is an inherently noisy process and that species might show both convergence and divergence in their traits as they drift over time. The community of coexisting mammals in Yotvata might therefore represent a nonrandom assembly of species that evolved more or less independently and that have subsequently been sorted into the community based on their traits. Alternatively, it is possible that traits have diverged after community assembly through evolutionary character displacement to reduce niche overlap. Although it is not straightforward to disentangle these two possible explanations in the current framework, one approach could be to explore the phylogenetic depth at which lineages show overdispersion. Species sorting on traits evolved in allopatry might reflect deeper evolutionary divergences, whereas evolutionary character displacement might be more apparent toward the tips of the tree, reflecting more recent evolutionary divergences in sympatry. Of course, it is likely that both mechanisms are important in shaping community structure (Cornwell and Ackerly 2009) and that different traits may demonstrate different patterns. For example, filtering may occur on traits that define species' fundamental abiotic niches, whereas competition acts on traits related to resource use (Ackerly et al. 2006).

6.3. CONCLUSION

The field of community phylogenetics has tended to focus on the assembly of species into communities given their functional traits and phylogenetic relationships. In this chapter, we have seen how phylogenetic methods can additionally provide insights into how species interactions might shape trait evolution and even illuminate the process of speciation itself. It is important, however, to remember that phylogeny is only informative when traits (including geographic distributions in the case of studies on mode of speciation) retain some phylogenetic structure (chapter 5). The lability of geographic ranges is well documented, and ancestral ranges are hard to estimate, as we still lack a good model of range evolution. The use of phylogenetic methods for inferring the geographic mode of speciation has, therefore, been questioned (Losos and Glor 2003). In our example of exploring range overlap in carnivores, we used only sister species pairs, therefore avoiding the necessity of reconstructing ancestral areas. Comparisons between clades descending from nodes nested deeper in the phylogeny might be more problematic.

In chapter 4 we emphasized the importance of selecting the appropriate null model in evaluating patterns of community structure. In this chapter we have shown that phylogeny itself can be used to generate a null distribution with which to test observed patterns. Here the null distribution is generated by evolving a trait down the branches of the phylogenetic tree, and we used a Brownian motion mode of evolution to approximate drift (i.e., where traits evolve independently in a manner analogous to a random walk). Rejection of this null distribution might indicate that trait evolution is shaped by processes other than drift. However, as for community phylogenetic structure, multiple processes other than competition might lead to rejection of the null hypothesis, and it is possible that phylogenetic error could bias us toward incorrectly inferring support for overdispersion, especially if pairwise phylogenetic distances separating taxa were greater in the true phylogeny. In addition, the underlying model of trait evolution even in the absence of biotic interactions might depart from Brownian expectations. For example, many traits might follow a constrained model of evolution, such that trait values have a tendency to return to a medial value (Harmon et al. 2010). As for all comparative approaches, it is important therefore that both the phylogeny and assumed model of evolution are good approximations of the true tree and evolutionary process. Advances in tree reconstruction methods have been rapid (chapter 2), and it is now possible to iterate models across a set of trees, such as a sample from a Bayesian posterior distribution (see, e.g., Arnold et al. 2010). It is now also straightforward to compare the fit of different evolutionary models in comparative data sets (e.g., Harmon et al. 2010) and simulate traits appropriately (chapter 5).

While phylogenetic approaches for detecting character displacement therefore have limitations, one important advantage is that they free us from the necessity to define the regional species pool. The use of null models in ecology has a long and sometimes acrimonious history (Gotelli 2001). We now have a much greater appreciation of the importance of model choice (chapter 4); nonetheless, delineating the relevant regional source pools remains as much an art as a science (Weiher and Keddy 1995b, De Bello 2012, Lessard, Belmaker, et al. 2012). Phylogeny provides one possible route to free us from this limitation.

CHAPTER 7

Phylogenetic Diversity across Space and Time

*The charm of history and its enigmatic lesson consist
in the fact that, from age to age, nothing changes
and yet everything is completely different.*

—*Aldous Huxley*

In this book we have, so far, examined phylogenetic patterns for single species sets, whether that set is a local community or some large-scale biota. However, diversity is distributed unequally in space and time, and reflects several underlying factors; these include environmental variability, dispersal limitation, and species interactions such as the presence of dominant competitors or keystone predators. These factors change across spatial and temporal gradients, and result in nonrandom diversity patterns.

Changes in diversity have been at the heart of ecological analyses from the very beginning, and explaining these changes was a challenge for early ecologists (King 1685, Blytt 1886, Spalding 1909, Warming 1909, Cowles 1911, Clements 1928). One of the first "modern" accounts of ecological community change was from Eugen Warming in the 1890s (see Clements 1928 for a review), who tried to discern the rules governing successional change. From one time or place to another, it was clear that plant composition changed, and the energy of many early plant ecologists was dedicated to describing assemblages in terms of the forms of species that occurred under certain conditions (Cowles 1911, Gleason 1922, Raunkiær 1934), whether "associations" of plants were regular occurrences, and what mechanisms generated them (Clements 1928, Lippmaa 1939, Watt 1947). During this period, the idea that the presence of species in an assemblage may be individualistic or stochastic (Spalding 1909, Gleason 1922) also began to emerge. In this developing view of plant community ecology several general rules became apparent: (1) the environment was of paramount importance in determining when and where species were found, and (2) species were limited in their abilities to reach new places.

This bipartite view of community assembly still dominates analyses today, with a number of robust statistical tools available to ecologists for exploring community patterns (e.g., Cottenie 2005); yet through the 1920s and 30s, analysis of community differences remained largely descriptive. Even though Jaccard (1901) published a method to measure the similarity among assemblages very early on, it wasn't until the 1940s that there was further development of metrics measuring compositional similarity (Sørensen 1948, Bray and Curtis 1957). Collectively, these measures quantified the similarity, or dissimilarity, of communities, depending on how elements where arranged; they were based on the numbers of unique and shared species between pairs of sites. The Jaccard index (S_J) can be represented as:

$$S_J = \frac{a}{(a+b+c)} \tag{7.1}$$

where a is the number of species shared in the two sites, b is the number of species unique to the first sample, and c is the number of species unique to the second sample. The Sørensen index (S_S) is quite similar to the Jaccard index but gives a greater weight to the shared species. Using the same notation, the Sørensen index is:

$$S_S = \frac{2a}{(2a+b+c)} \tag{7.2}$$

The final measure that we will briefly mention is the Bray-Curtis measure (BC), which is formulated as a dissimilarity, and to get the similarity we can simply calculate $1 - BC$. The dissimilarity is:

$$BC = 1 - \frac{2C_{i,j}}{(N_i + N_j)} \tag{7.3}$$

where $C_{i,j}$ is the sum of the absolute differences between each species abundance in the two samples and the N's are the summed abundances of the two samples. The Bray-Curtis is the abundance-weighted version of the Sørensen index, and changing abundances to presence-absence results in $BC = 1 - S_S$. To calculate these, we use the function `vegdist()` in the *vegan* library. To calculate the Jaccard index, using the example data in *vegan*, we do the following:

```
data(varespec)
vegdist(varespec, method="jaccard")
```

For the Sørensen:

```
vegdist(varespec, method="bray", binary=TRUE)
```

And now the Bray-Curtis (remember that it is analogous to the Sørensen when set to use presence-absence data):

```
vegdist(varespec, method="bray")
```

These functions return a distance matrix that can be then correlated with spatial or environmental distances. We will show how to implement equivalent similarity metrics using phylogenetic data (section 7.1.2) in place of species.

The other historical development that had a major impact on how ecologists examine changes in diversity was the emergence of the concept of the different spatial scales of diversity. Studying the vegetation of the Siskiyou Mountains, Robert H. Whittaker created a measure of turnover that quantified the relationship between local and regional diversity (Whittaker 1960). Whittaker referred to this turnover as beta diversity (β-diversity), whereas local diversity was termed alpha diversity (α-diversity), and diversity at the landscape scale gamma diversity (γ-diversity). While gamma diversity and alpha diversity can be calculated directly from species inventory data, beta diversity is often derived through additive ($β = γ - α$) or multiplicative partitioning ($β = γ/α$). However, it is important to remember that metrics of similarity, such as the Jaccard and Sørensen indices, can also be considered as capturing dimensions of beta diversity; the literature can be confusing on this concept (Anderson et al. 2011).

7.1. PHYLOBETADIVERSITY: MEASURING PHYLOGENETIC TURNOVER

In community ecology, studies examining compositional turnover, which we follow Whittaker in referring to as a dimension of beta diversity, are commonly employed to understand mechanisms structuring communities across spatial, environmental, and temporal gradients. In this vein, studies of phylogenetic turnover or phylogenetic beta diversity (phylobetadiversity) examine the combined influences of ecological processes (e.g., environmental filtering and biotic interactions) and evolution (especially speciation) on both local and regional patterns of biodiversity (Graham and Fine 2008). Broadly, phylobetadiversity is defined as the phylogenetic distances among communities (Graham and Fine 2008). This is in contrast to chapter 3, which focused on phylogenetic distances within communities. Throughout this section we will use β, meaning beta, interchangeably with phylobetadiversity; α, for alpha, to mean local, within patch phylogenetic diversity; and γ, or gamma, for regional diversity. The basic concept for these hierarchical diversity levels is shown in figure 7.1.

Generally, low phylobetadiversity corresponds to regions where distantly related taxa are widespread (low species beta diversity), or where close relatives spatially partition habitats (high species beta diversity) but more distantly related species are able to persist together (Graham and Fine 2008). We might see high turnover among closely related species for two reasons: first, because speciation is generally allopatric and speciation rates are high-producing specialized species in regions with ample habitat heterogeneity (Valente et al. 2009, Tucker et al. 2012); and second, because local competition and competitive exclusion limit coexistence by close relatives (e.g., Poffenroth and Matson 2007). (See chapters 3 and 6.) Given that the pattern of phylobetadiversity can be congruent or in conflict with the pattern of traditional species beta diversity, phylobetadiversity enriches our ability to infer ecological and evolutionary mechanisms that shape how diversity is distributed (Swenson 2011).

For evolutionarily distinct small-ranged species, we do not necessarily need to evoke a particular ecological mechanism for high phylobetadiversity, because randomly placing small-ranged species within a larger region will result in low overlap and high turnover. The interesting question might be why distinct species have small ranges. This could be because distinct species are either specialized or ill suited to modern environmental conditions, as opposed to speciose groups that thrive in modern environments (Simpson 1944). The alternative cause of high phylobetadiversity—low overlap of distant relatives—could be

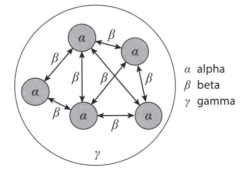

Figure 7.1. The hierarchical partitioning of diversity into within-habitat diversity (alpha), among-habitat diversity (beta), and total regional diversity (gamma).

different speciation histories or habitat or environmental preferences in heterogeneous landscapes. For example, in a landscape that transitions from forest to meadow, we should see high phylobetadiversity as entire clades are excluded (e.g., beech and oak trees in the Fagaceae) and other clades emerge in the open habitat (e.g., grasses [Poaceae] and asters [Asteraceae]). To untangle potential causes, we can examine how phylobetadiversity patterns are correlated with environmental and spatial gradients.

Like traditional species beta diversity, phylobetadiversity can be calculated a number of ways, and the number of phylobetadiversity metrics has been increasing. It is likely that a great many phylobetadiversity metrics are correlated with one another (Feng et al. 2012), but like species beta diversity we can group metrics into two general types (Tuomisto 2010): (1) diversity partitioning, and (2) pairwise distances.

7.1.1. Partitioning Phylogenetic Diversity

The partitioning of diversity is premised on quantifying our three types or scales of diversity (alpha, α; beta, β; and gamma, γ) (see fig. 7.1). Alpha diversity is usually measured as the average diversity within patches ($\bar{\alpha}$), whereas gamma diversity is the diversity represented by the regional species pool. It can be quite difficult to determine what should constitute the regional species pool (Srivastava 1999), and however we decide upon a regional pool there will be important repercussions for subsequent analyses and inference (see chapter 4). As we mention above, there are two common approaches to calculate beta diversity from alpha and gamma. The first one is a multiplicative approach, calculated as:

$$\beta = \frac{\gamma}{\bar{\alpha}} \text{ or else } \beta = \frac{(\gamma - \bar{\alpha})}{\bar{\alpha}} \tag{7.4}$$

The second formulation was the original concept of beta diversity developed by Whittaker (1960), but these two measures essentially quantify how many times species composition changes along a gradient or across habitats. The other approach is referred to as "additive" or "absolute" and is calculated as:

$$\beta = \gamma - \bar{\alpha} \tag{7.5}$$

In this formulation, introduced independently several times (Veech et al. 2002) but popularized by Lande (1996), beta is the average number of species not found within a patch, and is preferred by some because it fits concepts of additive variance partitioning better. It is important to note that despite arguments and traditions that support the use of one form of beta partitioning over the other (Veech et al. 2002, Tuomisto 2010), they are obviously intimately linked and are strongly correlated (fig. 7.2). Thus, the choice of partitioning method is not likely to be that consequential, and for our examples we will use the additive method.

It seems that it should be straightforward to use a partitioning equation to calculate phylobetadiversity for our phylogenetic metric of interest. However, to partition diversity our metric must not violate the following condition: gamma must be greater than or equal to alpha, as this would otherwise return negative betas. There are several metrics that do not fulfill this condition. For example, MPD_{gamma} could conceivably be less than MPD_{alpha}. Thus only phylogenetic metrics that sum across species or branches should be used. The most obvious metric for diversity partitioning is PD, because PD_{gamma} will always be greater than or equal to PD_{alpha}. Using the additive approach, PD_{alpha} is the average PD within patches or

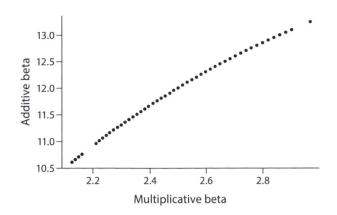

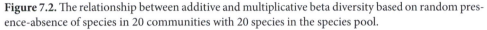

Figure 7.2. The relationship between additive and multiplicative beta diversity based on random presence-absence of species in 20 communities with 20 species in the species pool.

plots, PD_{gamma} is the total branch lengths or evolutionary history represented in the species pool, and PD_{beta} is the average amount of PD not found in a patch or plot. Here is a function to calculate partitioned PD:

```
phylo.part<-function(phy,com){
  PDt<-sum(phy$edge.length)
  PDa<-mean(pd(com,phy)[,1])
  out<-data.frame(PDt=PDt,PDa=PDa,
    PD.bet.add=PDt-PDa,
    PD.bet.whit=(PDt-PDa)/PDa,
    PD.bet.mult=PDt/PDa)
  return(out)
}
```

Let's examine partitioned diversity values for the Jasper Ridge data used in chapter 3:

```
library(picante)

phylo.part(j.tree,j.com)
```

This returns:

PDt	PDa	PD.bet.add	PD.bet.whit	PD.bet.mult
1508.402	778.6387	729.763	0.9372293	1.937229

And this tells us that the species pool contains over 1,508 million years of evolutionary distance (`PDt`), with an average of more than 778 million years within plots (`PDa`). On average there is almost 730 million years not represented in individual plots (`PD.bet.add`), or almost an equal amount of PD not within plots. The `PD.bet.whit` and `PD.be.mult` return Whittaker's original beta diversity formulation and the multiplicative version, respectively.

Other phylogenetic measures can be used in a partitioning a framework—for example, some of the entropic measures that we include in table 3.3 (Hardy and Senterre 2007)—but the behavior of metrics should be considered carefully. The diversity partitioning calculations generate single values for an entire region and can only be used in analyses that compare regions. For analyses that test hypotheses about turnover of phylogenetic relationships within regions, we are better off using a pairwise distance approach.

7.1.2. Pairwise Distance Matrices

For analyzing the diversity patterns among plots or patches, the distance matrix approach has become very popular. For ecophylogenetic analyses this means that several different distance matrices should be computed (see the next section for a description of types of analyses). First and foremost will be a site-by-site spatial distance matrix that simply contains geographical distances between patches. Robust inferences will result from the inclusion of an environmental matrix that contains environmental distances calculated from any number of approaches, including Euclidean distances, principle components analysis, and so forth. These two matrices are then used to explain variation in diversity differences or similarities among patches (fig. 7.3).

Calculating patch pairwise diversity distances is a critical step for these analyses and there are multitudes of ways to do this (Jost et al. 2011). Compositional distances, such as the Jaccard or Sørensen indices, compare the number of shared and unshared species, and there are phylogenetic analogues to these types of measures. Here we will discuss ten phylogenetic metrics that calculate pairwise distances among communities (table 7.1).

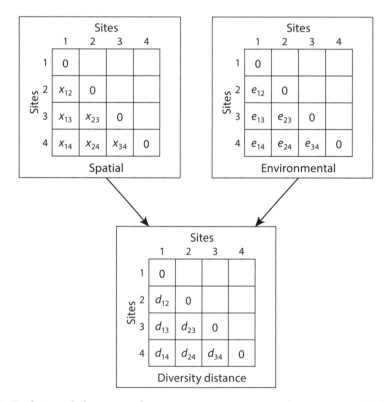

Figure 7.3. Explaining differences in diversity among sites requires three matrices: (1) the diversity matrix (where cell values are richness, PD, etc.); (2) a spatial matrix (where cell values are spatial distances between sites, e.g., km); and (3) an environmental matrix (where cell values are environmental distances calculated from multivariate analysis of measured abiotic variables).

TABLE 7.1

Common parameters used:
- k_1, k_2 = Habitat 1, Habitat 2
- S = Number of species
- i, j = Species i and j
- f = The relative abundance of a species
- N = Number of nodes
- e = Edge e
- $\bar{d}_{ik_2}$ = The average pairwise phylogenetic distance between species i in community 1 to all species in community 2
- $\bar{d}_{jk_1}$ = The average pairwise phylogenetic distance between species j in community 2 to all species in community 1
- $\lambda_1, \lambda_2, \lambda_{12}$ = The edges found in community 1, 2, and both, respectively
- min d_{ik_2} = The nearest phylogenetic distance between species i in community 1 to its closest relative in community 2
- min d_{jk_1} = The nearest phylogenetic distance between species j in community 2 to its closest relative in community 1
- $\bar{d}_{k_1}$ = Average pairwise distance in community 1
- $\bar{d}_{k_2}$ = Average pairwise distance in community 2
- S_{ek1} = The number of species that descend from edge e in community 1
- λ_e = Length of edge e

Metric	Equation	Citation
Pairwise phylogenetic dissimilarity	$$D_{pw} = \frac{\sum_{i=1}^{Sk_1} \bar{d}_{ik_2} + \sum_{j=1}^{Sk_2} \bar{d}_{jk_1}}{2}$$	(Swenson 2011)
Abundance-weighted pairwise phylogenetic dissimilarity	$$D'_{pw} = \frac{\sum_{i=1}^{Sk_1} f_i \cdot \bar{d}_{ik_2} + \sum_{j=1}^{Sk_2} f_j \cdot \bar{d}_{jk_1}}{2}$$	(Swenson 2011)
Pairwise nearest neighbor dissimilarity	$$D_{nn} = \frac{\sum_{i=1}^{Sk_1} \min d_{ik_2} + \sum_{j=1}^{Sk_2} \min d_{jk_1}}{2}$$	(Swenson 2011)
Abundance-weighted pairwise nearest neighbor dissimilarity	$$D'_{nn} = \frac{\sum_{i=1}^{Sk_1} f_i \cdot \min d_{ik_2} \sum_{j=1}^{Sk_2} f_j \cdot \min d_{jk_1}}{2}$$	(Swenson 2011)
Weighted average difference between two individuals	$$D_{Rao's} = \sum_{i=1}^{Sk_1} \sum_{j=1}^{Sk_2} d_{ij} \, f_{ik_2} \, f_{jk_1}$$	(Rao 1982, Hardy and Senterre 2007, Swenson 2011)
$D_{Rao's}$ standardized by differences in patch MPD	$$H_{Rao's} = \frac{2 \cdot D_{Rao's}}{\left(\sum_{i=1}^{Sk_1} f_i \cdot \bar{d}_{k_1} - \sum_{j=1}^{Sk_2} f_i \cdot \bar{d}_{k_2} \right)}$$	(Swenson 2011)
Phylogenetic Sørensen's index	$$PhyloSor = \frac{2 \cdot \sum \lambda_{12}}{\sum \lambda_1 + \sum \lambda_2}$$	(Bryant et al. 2008)
Unique fraction	$$UniFrac = \sum_{e}^{N} \lambda_e \cdot \left| \frac{S_{ek_1} \cdot S_{ek_2}}{S_{k_1} \cdot S_{k_2}} \right|$$	(Lozupone and Knight 2005, Swenson 2011)

Swenson (2011) groups these metrics into those that are terminal (*Phylosor, UniFrac, D_{nn} and D'_{nn}*) and those that are basal (*D_{pw}, D'_{pw}, $D_{Rao's}$ and $H_{Rao's}$*). Terminal metrics include those that are sensitive to turnover near the tips of the trees, whereas basal metrics are sensitive to turnover deeper in the phylogeny. All of these metrics stem from other measures. *Phylosor* is the phylogenetic counterpart to the Sørensen index, and *UniFrac* is analogous to the Jaccard index. The rest of the measures in table 7.1 have within-community counterparts: MPD and MNTD for D_{pw} and D_{nn}, respectively, and Rao's quadratic for the two Rao measures.

Studies have shown that the two groupings of phylobetadiversity (terminal and basal) do show differing behaviors across spatial and environmental gradients (Swenson 2011, Jin et al. 2015). Further, measures within groups tend to be correlated with one another, and measures between the two groups have been shown to be uncorrelated or only weakly correlated (Swenson 2011). Let's examine this using the Jasper Ridge data that we loaded earlier. We will use *picante* to calculate these measures of phylobetadiversity, but there are other packages (e.g., *betapart*)that also calculate some of these measures, and additional ones that quantify turnover and nestedness as separate quantities (Baselga 2010).

Here, we will calculate all the metrics except D'_{nn} and D'_{pw}, because these abundance-weighted versions show the same basic behavior as their unweighted counterparts. For all the metrics except the two Rao's measures, the functions return a triangular distance matrix (e.g., fig. 7.3). So let's calculate those first using `picante` functions:

```
beta_Dpw<-comdist(j.com,cophenetic(j.tree))
beta_Dnn<-comdistnt(j.com,cophenetic(j.tree))
beta_ps<-phylosor(j.com,j.tree)
beta_uf<-unifrac(j.com,j.tree)
```

Now for Rao's, there is a convenient function in *picante*, called `RaoD()`, which calculates both within and among estimates and returns a list with a number of elements, only some of which are of interest to us here. Let's run the function:

```
rao.tmp<-raoD(j.com,j.tree)
```

You can use a number of different functions to see what the list contains without actually calling the list—if it is a large data set, we would not want to call the full list, just a small part of it. For example, we can use `attributes()` or `names()`:

```
names(rao.tmp)
[1] "Dkk" "Dkl" "H" "total" "alpha" "beta" "Fst"
```

Here, `Dkk` is within community Rao's D, `Dkl` is Rao's D among communities, `H` is the among community diversity excluding the within diversity, and the next three are the additive components: total (gamma), alpha, and beta. We haven't yet discussed `Fst`, but it is equivalent to the fraction of total diversity partitioned among communities (Hardy and Senterre 2007):

$$F_{st} = \frac{D_\gamma - D_\alpha}{D_\gamma} \qquad (7.6)$$

This returns a single value, much like our other partitioned beta diversities.

To extract `Dkl` and `H` from our list we need to turn them into `dist` objects, matching with those produced by our other distance function in *picante*:

```
beta_rD<-as.dist(rao.tmp$Dkl)
beta_rH<-as.dist(rao.tmp$H)
```

To ensure that this is actually a **dist** object use:

```
class(beta_rD)
[1] "dist"
```

These beta diversities are all triangular distance matrices with the same dimensions, and we can simply plot them against one another. First we can see how the three "terminal" metrics are related to one another:

```
Par(mfrow=c(1,3),cex.lab=1.3)
plot(beta_Dnn,beta_ps,xlab="Dnn",ylab="PhyloSor",pch=19)
plot(beta_Dnn,beta_uf,main="Terminal
        metrics",xlab="Dnn",ylab="UniFrac",pch=19)
plot(beta_ps,beta_uf,xlab="PhyloSor",ylab="UniFrac",pch=19)
```

This produces figure 7.4. It is clear that there are similarities between these measures. **PhyloSor** is negatively related to the others because it is actually a similarity measure, whereas the others are difference measures.

To make **PhyloSor** a difference measure we simply subtract the **PhyloSor** values from 1. Figure 7.5 shows the relationship between **Dnn** and 1 – **PhyloSor**, showing that it is analogous to the relationship between **Dnn** and **UniFrac** shown in figure 7.4. This example highlights the need to understand your metric—whether it is a similarity or difference measure—and how to compare between them. If we were analyzing the relationship between spatial distance and beta diversity, we would come to very different conclusions about the decay of similarity if we erroneously used a similarity measure instead of a difference measure:

```
beta_ps_D<-1-beta_ps
```

```
plot(beta_Dnn,beta_ps_D,xlab="Dnn",ylab="1-PhyloSor",pch=19)
```

Next, we will examine how the basal metrics compare to one another. We plot these measures using a set of commands similar to the previous example, and we produce figure 7.6:

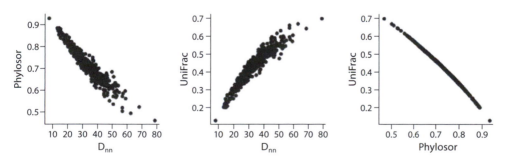

Figure 7.4. Relationships between the terminal phylobetadiversity metrics.

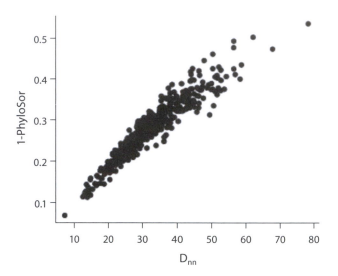

Figure 7.5. The relationship between 1 – Phylosor and D_{nn}.

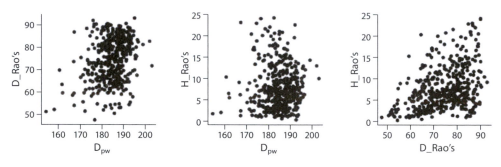

Figure 7.6. The relationships between the basal phylobetadiversity metrics.

```
par(mfrow=c(1,3),cex.lab=1.3)
plot(beta_Dpw,beta_rD,xlab="Dpw",ylab="D_Rao's",pch=19)
plot(beta_Dpw,beta_rH,main="Basal
        metrics",xlab="Dpw",ylab="H_Rao's",pch=19)
plot(beta_rD,beta_rH,xlab="D_Rao's",ylab="H_Rao's",pch=19)
```

Here it is clear that the basal measures are not as tightly correlated to one another as the terminal measures are to each other. This is because, in part, Dpw is not abundance weighted and the Rao measures include abundance. So let's redraw figure 7.6 using abundance-weighted Dpw (D'pw):

```
beta_Dpw_ab<-comdist(j.com,cophenetic(j.tree),abundance.
        weighted=TRUE)

par(mfrow=c(1,3),cex.lab=1.3)
plot(beta_Dpw_ab,beta_rD,xlab="D'pw",ylab="D_Rao's",
        pch=19)
```

```
plot(beta_Dpw_ab,beta_rH,main="Basal
      metrics",xlab="D'pw",ylab="H_Rao's",pch=19)
plot(beta_rD,beta_rH,xlab="D_Rao's",ylab="H_Rao's",pch=19)
```

In figure 7.7, it is now obvious that D'pw is just Rao's D multiplied by 2. Rao's H remains quite divergent from the other two measures. This is because Rao's H standardizes the beta measure by the MPD of communities, which drives much of the variation in the other two measures. Thus Rao's H can be thought of as a measure with a stronger imprint of abundance. Finally, we can ask how terminal and basal metrics compare to one another by plotting Dnn against Dpw (fig. 7.8). We will also calculate the correlation coefficient:

```
cor(beta_Dpw,beta_Dnn)
[1] 0.1514515
plot(beta_Dpw,beta_Dnn,main="Terminal vs.
      basal",xlab="Dpw",ylab="Dnn",pch=19)
```

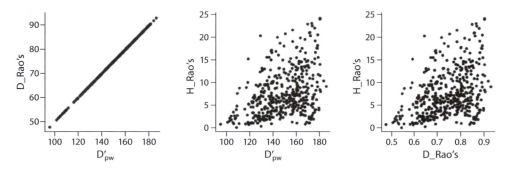

Figure 7.7. The relationships between the basal abundance-weighted phylobetadiversity metrics.

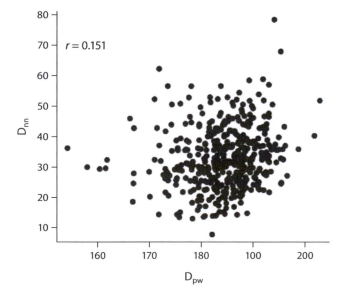

Figure 7.8. Terminal and basal phylobetadiversity metrics are uncorrelated.

It is clear that `Dnn` and `Dpw` are only very weakly correlated, which supports the need to carefully consider choice of a metric.

The distance matrices calculated above fit naturally into analyses that examine the influence of interpatch distances and environmental differences on diversity (fig. 7.3). It is straightforward to build a spatial distance matrix (e.g., km) between patches; this is usually measured from the nearest patch edge or patch centers. Constructing a site-by-site environmental distance matrix requires additional analyses and decisions. For these examples we will again use the Jasper Ridge data as our platform. Because we do not have information on the spatial and environmental attributes for the Jasper ridge plots, for illustration we will simulate random data. The simplest way to construct an environmental distance matrix would be calculate the Euclidean distances (ED) between patches as, for example, between patch 1 and 2:

$$ED_{1,2} = \left(\sum_{v=1}^{X} (X_{v1} - X_{v2})^2 \right)^{1/2} \tag{7.7}$$

Here X is the value of environmental variable v of V variables. Let's create random environmental data with four environmental variables, which are drawn from different distribution types:

```
env<-data.frame(row.names=row.names(j.com),
    env1=abs(rnorm(nrow(j.com),22,5)),
    env2=abs(rexp(nrow(j.com),3)),
    env3=abs(rgamma(nrow(j.com),4,3)),
    env4=abs(rnorm(nrow(j.com),4)))
```

This produces a data frame with the same number of sites as in the Jasper Ridge data (`nrow(j.com)`). The function `abs()` returns the absolute values so we do not get negative values, and each of the functions beginning with "r" is a random draw from a distribution (normal, exponential, gamma and normal, respectively). Now let's create a Euclidean distance matrix:

```
euc<-dist(env,method="euclidean",p=2)
```

The problem with Euclidean distances is that positively correlated environmental variables inflate patch differences. So we will use principal components analysis (PCA) to help deal with correlated variables. We will use `prcomp()` to perform the PCA, and then `scores()` to create a data frame with the first three PCA axes:

```
pc <-prcomp(env, scale = TRUE)
pc <-scores(pc, display = "sites", choices = 1:3)
```

We can see how the communities relate to one another in PCA space by plotting the community labels for the first two axes:

```
plot(pc,typ="n")
text(pc[,1],pc[,2],labels=rownames(pc))
```

This returns figure 7.9. From this example we can see that communities J18, J39, and J26 are environmentally distinct relative to the other communities. Further, there is a group of communities in the center showing high similarity with one another. It is important to remember that these associations have occurred by chance because we generated random environmental data to associate with each community, and every run of our code will produce different output.

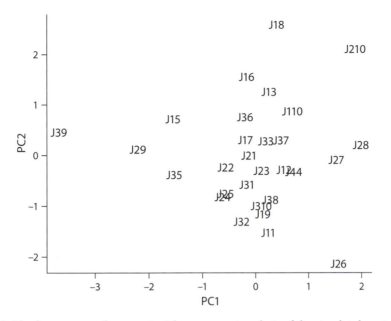

Figure 7.9. The first two axes from a principle component analysis of the simulated environmental data for Jasper Ridge.

Now let's create the Euclidean distance matrix from the PCA components:

```
euc2 <-dist(pc, method = "euclidean")
```

We can create a random spatial distance matrix for the Jasper data. We generate this here by converting one of our beta matrices (in our case, Rao's *H*) into a new measure by multiplying the values by numbers randomly drawn from a normal distribution:

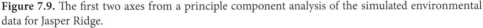

```
space<-abs(beta_rH*rnorm(length(as.vector(beta_rH)),25,10))
```

We could have simply drawn our spatial distance matrix at random from some distribution, as we did for our environmental distances, but by multiplying Rao's *H* we introduce spatial covariance in our matrix, allowing us to better evaluate the power of our methods for detecting signal in the data.

Now that we have our spatial and environmental distance matrices, we can ask if they explain beta diversity patterns. Let's look at how space and environment appear to correlate with the different betas by graphing the relationships. First let us look at space:

```
par(mfrow=c(2,3),cex.lab=1.3)
plot(space,beta_Dnn,xlab="",ylab="Dnn",pch=19)
plot(space,beta_ps_D,xlab="",ylab="1-PhyloSor",pch=19)
plot(space,beta_uf,xlab="",ylab="UniFrac",pch=19)
plot(space,beta_Dpw,xlab="",ylab="Dpw",pch=19)
plot(space,beta_rD,xlab="Distance (m)",ylab="Rao's
        D",pch=19)
plot(space,beta_rH,xlab="",ylab="Rao's H",pch=19)
```

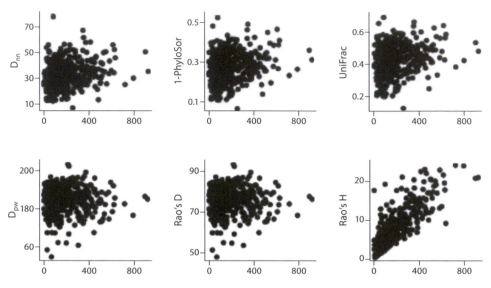

Figure 7.10. Relationships between spatial distances and phylobetadiversity measures.
Spatial distance (m)

This generates figure 7.10. There appears to be some relationships, especially with Rao's *H* (remember how our data was simulated before reading ecological explanations into these relationships).

We can do the same thing for the environmental distances:

```
par(mfrow=c(2,3),cex.lab=1.3)
plot(euc2,beta_Dnn,xlab="",ylab="Dnn",pch=19)
plot(euc2,beta_ps_D,xlab="",ylab="1-PhyloSor",pch=19)
plot(euc2,beta_uf,xlab="",ylab="UniFrac",pch=19)
plot(euc2,beta_Dpw,xlab="",ylab="Dpw",pch=19)
plot(euc2,beta_rD,xlab="Environmental distance",ylab="Rao's
        D",pch=19)
plot(euc2,beta_rH,xlab="",ylab="Rao's H",pch=19)
```

This produces figure 7.11. Here it is less clear if any of the beta diversities are correlated with environment.

Now, how do we actually analyze these relationships? We can't use traditional correlation or regression analysis because the data points in figures 7.10 and 7.11 are not independent from one another as individual sites are partnered in numerous distances. Given this nonindependence, we are required to use analyses specifically designed for distances. We will briefly discuss three such methods: (1) Mantel tests, (2) Procrustes analysis, and (3) generalized distance modeling (GDM). Regardless of the particular statistical approach, our goal is to assess the relative contribution of space and environment to variation in phylobetadiversity. In these examples we are using an independent distance matrix (of environment) to explain variation in phylogenetic distances. If we had a grouping variable as our independent variable, we could use a permutation-based analogue to an ANOVA using functions like `anosim()` or `adonis()` in the `vegan` package.

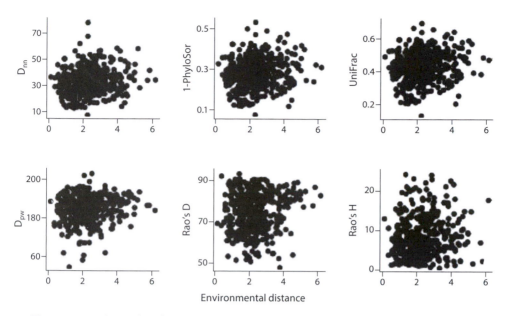

Figure 7.11. Relationships between environmental distances and phylobetadiversity metrics.

Mantel tests are analogous to correlations but are based on permutation tests to assess significance. We will use the `mantel()` function in the **vegan** library. First let's examine how well the different phylobetadiversities are correlated with geographical distance. The Mantel test looks much like a regular correlation call, but we need to specify the number of permutations. We will use 999:

```
m1<-mantel(beta_Dnn,space, permutations=999)
m2<-mantel(beta_ps_D,space, permutations=999)
m3<-mantel(beta_uf,space, permutations=999)
m4<-mantel(beta_Dpw,space, permutations=999)
m5<-mantel(beta_rD,space, permutations=999)
m6<-mantel(beta_rH,space, permutations=999)
```

Let's look at the results of one of these; to do this we simply type the name we assigned to store the model results (m6):

```
Mantel statistic based on Pearson's product-moment
        correlation

Call:
mantel(xdis = beta_rH, ydis = space)

Mantel statistic r: 0.8201
    Significance: 0.000999

Upper quantiles of permutations (null model):
  90%    95%   97.5%    99%
0.127  0.156  0.187  0.224

Based on 1000 permutations
```

Here we can see that Rao's H is significantly correlated with space ($P < 0.001$) and the positive r statistic means that communities further apart are more different, according to Rao's H. We can call each of the saved Mantel outputs (m1, m2, etc.) to see the results for the different metrics, and we will see that space is a significant predictor ($P < 0.05$) for all the phylobetadiversity measures except for D_{pw} ($P = 0.481$). Recall that we generated the spatial distance matrix by multiplying Rao's H by a number randomly drawn from a normal distribution; we thus generated strong spatial covariance in our matrix, which our Mantel tests identify correctly. What about environment? Let's run Mantel tests using the environmental distances:

```
e1<-mantel(beta_Dnn,euc2, permutations=999)
e2<-mantel(beta_ps_D,euc2, permutations=999)
e3<-mantel(beta_uf,euc2, permutations=999)
e4<-mantel(beta_Dpw,euc2, permutations=999)
e5<-mantel(beta_rD,euc2, permutations=999)
e6<-mantel(beta_rH,euc2, permutations=999)
```

If we look at the output from these tests we will see that none of the basal phylobetadiversity measures are correlated with environmental distance, but in our example the terminal metrics do appear to be correlated with environmental distance ($P \leq 0.05$ for D_{nn}, *PhyloSor*, and UniFrac). We should be cognizant of the fact that environmental differences will often be correlated with spatial distances (in our case they are not because of the random data), making inference difficult. We want to distinguish between two hypotheses: (1) phylobetadiversity is correlated with space, independent of the environment, which reflects dispersal limitation; versus (2) phylobetadiversity is correlated with environmental differences, independent of spatial distance, which reflects environmental filtering. To do this, we can use partial Mantel tests. Partial Mantel tests allow us to derive the correlation between beta and environment given space ($r_{\beta S|E}$), or the correlation between beta and space given the environment ($r_{\beta E|S}$). Partial Mantel tests may be especially important if we find that both predictors are significant on their own. We will show an example of both partial Mantel tests here, looking for the effects of space and environment on D_{nn}:

```
mantel.partial(beta_Dnn,euc2,space)
```

This returns:

```
Partial Mantel statistic based on Pearson's product-moment
    correlation

Call:
mantel.partial(xdis = beta_Dnn, ydis = euc2, zdis = space)

Mantel statistic r: 0.1477
    Significance: 0.087

Upper quantiles of permutations (null model):
  90%    95%  97.5%   99%
0.131  0.177  0.213 0.264

Permutation: free

Number of permutations: 999
```

This reveals that the effect of environment on beta after accounting for the effect of space (included as the last argument in the function) is marginally nonsignificant ($P = 0.087$). Now we can test for the effect of space given the environment:

```
mantel.partial(beta_Dnn,space,euc2)
```

This returns:

```
Partial Mantel statistic based on Pearson's product-moment
     correlation

Call:
mantel.partial(xdis = beta_Dnn, ydis = space, zdis = euc2)

Mantel statistic r: 0.2036
    Significance: 0.012

Upper quantiles of permutations (null model):
   90%     95%   97.5%    99%
 0.100   0.130   0.167  0.204

Permutation: free

Number of permutations: 999
```

This reveals that space is a significant predictor of beta, even after accounting for the effect of environment.

There are well known issues with Mantel tests, such as low statistical power (Peres-Neto and Jackson 2001, Legendre and Fortin 2010) and spatial autocorrelation biases (Guillot and Rousset 2013); in addition, partial Mantel tests do not allow for the analysis of individual cases (Peres-Neto and Jackson 2001). These problems have led researchers to explore alternative approaches. One of these is the Procrustes[1] superimposition method (Jackson 1995), which is also a permutation method to assess the correlation between two distance matrices. Procrustes is a complex statistical routine that rescales and rotates data to minimize the sum of squares in multivariate space (Peres-Neto and Jackson 2001). We will run through a Proscrustes example using Rao's H. The input for Procrustes is a data matrix with some number of axes from multivariate analysis (e.g., principal component analysis, principal coordinate analysis, nonmetric multidimensional scaling, etc.). Since our Rao's H is a distance matrix, we need to perform a principal coordinate analysis (PCoA) on the distance matrix, and we will take the first three axes:

```
p_rH<-pcoa(beta_rH)
p_rH<-p_rH$vectors[,1:3]
```

Our spatial distances are contained in a distance matrix as well, so we will need to use PCoA again, also taking the first three axes:

```
p_space<-pcoa(space)
p_space<-p_space$vectors[,1:3]
```

And now we run the `protest()` function in the `vegan` package:

```
pro.s<-protest(p_rH,p_space)
```

1. Procrustes is a mythological figure from ancient Greece, who made people fit an iron bed by either stretching them or cutting off their legs. Sounds morbid, but the statistical logic should be obvious.

This returns the following output (note that rerunning the creation of the random spatial matrix will result in different values):

```
Call:
Protest(X = p_rH, Y = p_space)

Procrustes Sum of Squares (m12 squared):          0.3716
Correlation in a symmetric Procrustes rotation:   0.7927
Significance:  0.001
Based on 999 permutations
```

The Procrustes correlation ($r = 0.793$) is similar to the Mantel correlation ($r = 0.820$), revealing that in this case Procrustes and Mantel analyses detect significant correlations similarly. In Procrustes, the correlation coefficient is calculated from the sum of squares as:

$$r = \sqrt{1 - m^2} \tag{7.8}$$

For the environmental data, we can use the "pc" data object created earlier from the PCA of the environmental variables and run the analysis again:

```
pro.e<-protest(p_rH,pc)
```

This results in:

```
Call:
Protest(X = p_rH, Y = pc)

Procrustes Sum of Squares (m12 squared):        0.9341
Correlation in a symmetric Procrustes rotation: 0.2567
Significance:    0.459
Based on 999 permutations
```

Although we have a higher correlation coefficient than for the Mantel test ($r = 0.257$ for Procrustes versus 0.169 for Mantel), neither result is statistically significant ($P = 0.459$ and $P = 0.052$) and so we cannot read too much into these coefficients. We can also perform a partial Procrustes analysis following Peres-Neto and Jackson (2001) if we want to test the relationship between Rao's H and environment controlling for space. First we remove the effect of space on Rao's H by regressing the PCoA of Rao's H against the PCoA of space. Then we remove the effect of space on environment by regressing the PCA from the environmental variables against the PCoA of space. We can then use the residuals from both of these analyses in the Procrustes test:

```
pp.s<-lm(p_rH~p_space)
pp.s.res<-pp.s$residuals

pp.e<-lm(pc~p_space)
pp.e.res<-pp.e$residuals

pro.pp<-protest(pp.e.res,pp.s.res)
```

This returns:

```
Protest(X = pp.s.res, Y = pp.e.res)

Procrustes Sum of Squares (m12 squared):        0.9083
Correlation in a symmetric Procrustes rotation: 0.3028
```

```
Significance:   0.223
Based on 999 permutations
```

This tells us that the environment does not influence Rao's H once we control for the effect of space. We could have equally removed the effect of environment on Rao's H and spatial distance, as we did for the partial Mantel tests.

The final approach is to use generalized dissimilarity modeling (GDM; Ferrier et al. 2007) as applied to phylogenetic data (Rosauer et al. 2014). The GDM approach is a statistical model used to analyze species turnover among plots that is a result of spatial and environmental distances (Ferrier et al. 2007). Rosauer et al. (2014) extended the GDM to analyze phylogenetic turnover (using 1-PhyloSor). For those that are accustomed to using linear models, the GDM approach may be more intuitive than other approaches, but at the time of writing this book, the R code to implement phylo-GDM is not yet widely available, so we do not review it here.

All three approaches (Mantel, Procrustes, and GDM) are useful for testing hypotheses about the relative importance of spatial distance and environmental gradients in shaping community differences (e.g., Cottenie 2005). However, with phylogenetic information, we have added an extra layer to the analyses and we can thus make additional inferences. If we run these analyses on terminal and basal phylogenetic turnover metrics, along with taxonomic ones, we can assess whether there is a phylogenetic underpinning of species turnover and the relative depth of this phylogenetic structuring (Jin et al. 2015). However, we shouldn't compare the absolute magnitude of turnover values, because, as discussed by Jin et al. (2015), taxonomic turnover will always be higher than phylogenetic turnover. The reason is that the taxonomic distance measures are essentially calculated with a star phylogeny (see chapter 3), and therefore adding or subtracting a species from an assemblage involves adding or removing a hypothetical edge with the distance equivalent to the total root-to-tip distance. Phylogenetic turnover most often adds or subtracts edges much less than the root-to-tip length.[2]

Rather than comparing the magnitude of turnover among metrics, we should instead assess the significance of the associations with space and environment. If basal metrics show significant association with space (after accounting for environment), we might infer that deeper lineages or clades are spatially congruent and turnover together, perhaps suggesting that evolutionarily distinct species or groups of close relatives are dispersal limited or influenced by some landscape barrier. If we observe a significant association with environment (after accounting for space), we might infer that close relatives respond to the environment similarly, and that turnover across the environmental gradient involves clusters of closely related species.

If terminal metrics show significant associations, we might infer that most of the turnover is among closely related species and that internal edges tend to be found everywhere. This pattern could result from competitive exclusion among close relatives and we could complement turnover analyses with further analyses described in chapters 3 and 6 in order to understand how relatedness influences coexistence.

7.2. THE INFLUENCE OF SPATIAL SCALE ON PHYLOGENETIC PATTERNS

Intricately linked with the scale of observation of ecological patterns is the dominant mechanism structuring communities, and changes in scale often correspond to changes

---

2. We can of course use null models, as described in chapter 4, to evaluate whether phylobetadiversity is greater than expected given observed taxonomic turnover.

in the important mechanisms (Ricklefs and Schluter 1993, Peterson and Parker 1998). At small scales, biotic interactions (especially competition and facilitation) are thought to dominate; while at larger scales, habitat heterogeneity, dispersal limitation, stochastic local assembly and priority effects, and underlying environmental gradients influence where organisms will occur (Ricklefs and Schluter 1993, Chase and Leibold 2002, Cadotte 2006).

While chapter 3 dealt with local scale patterns and section 7.1 with beta diversity, these types of analyses assume that observations at those scales are concordant with the mechanisms being tested; but often they are not. The field of metacommunity ecology is largely built on the premise that different structuring mechanisms dominate at different spatial scales (Holyoak et al. 2005). However, there is an insidious side to scale dependency, which is an underappreciated issue influencing observed patterns. For example, there are potentially many reasons why one study may show phylogenetic overdispersion, and another may show clustering. It could be that the dominant mechanisms are fundamentally different, or it could be that there is a mismatch in the scale of observation and the likelihood of observing particular patterns. Further, how does one compare ecophylogenetic patterns from a 10 m^2 forest plot of trees to a 1 m^2 plot of perennial herbs? Perhaps the more fundamental question to ask is whether there are commonly observed influences of scale on community phylogenetic patterns.

7.2.1. Scale Dependency of Phylogenetic Patterns

While community ecology, and especially metacommunity ecology, incorporates processes operating at different scales into broader theories, ecophylogenetics offers unique insights into scale-dependent community patterns. There have been a plethora of studies that have examined phylogenetic patterns at different scales, with surprisingly robust generalizations emerging. All else being equal, studies have either regularly (but not invariably) observed overdispersion at small spatial scales (Cavender-Bares et al. 2006, Swenson et al. 2007, Bino et al. 2013, De Bello et al. 2013, Muenkemueller et al. 2014, Parmentier et al. 2014) and clustering at large scales (Cavender-Bares et al. 2006, Cadotte, Hamilton, et al. 2009, Kraft and Ackerly 2010, De Bello et al. 2013, Pearse et al. 2013, Parmentier et al. 2014). This shift in phylogenetic pattern with scale is often thought to reflect a switch from biotic interactions (e.g., competition, pathogen sharing, etc.) at small scales to abiotic filters at large scales that are driven by habitat heterogeneity and the inclusion of alternative environmental regimes (fig. 7.12). These scale-dependent mechanisms make intuitive sense because antagonistic interactions are strongest between organisms near one another—especially for sessile organisms (e.g., plants), which are most often examined in ecophylogenetic analyses. Conversely, at larger scales, habitat heterogeneity means that patchworks of habitats offer opportunities to different kinds of organisms that would not normally persist together in a single habitat type but that have been filtered on similar climatic tolerances.

We might be inclined to think that we should expect more phylogenetic diversity (greater phylogenetic distances separating taxa) at larger spatial scales, since there should be more habitat heterogeneity. However, it is important to remember that testing for overdispersion or clustering is predicated on the availability of some larger-scale species pool (chapters 3 and 4). The species pool for large regions can be an entire continent. Thus, we might find clustering at these scales because we are examining biogeographical

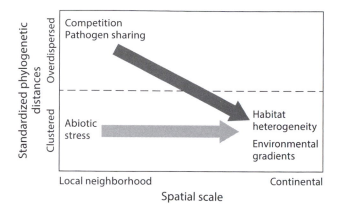

Figure 7.12. The hypothesized change in community phylogenetic patterns across spatial scale. At extremely large scales (e.g., continental), we should see signatures of phylogenetic clustering. Depending on the dominant structuring mechanisms, small-scale patterns could be overdispersed or clustered (see chapter 3).

and macroevolutionary processes, and any large region will likely have specific environmental regimes or past speciation events that produce phylogenetically clustered biotas (see chapter 8). A most extreme example would be if we study the flora of Arizona, USA, and use as the species pool all of the plants found in western North America. The prevalence of a few distinct clades (e.g., cacti) would undoubtedly generate an observation of clustering.

The general trend outlined in figure 7.12 is supported by patterns of non-native species. Non-natives often serve as a powerful natural experiment for testing ecological hypotheses (Cadotte et al. 2006, Sax et al. 2007). For phylogenetic patterns, non-native species provide researchers with the opportunity to observe recent dynamic assembly mechanisms. Carboni et al. (2013) observed that coastal plant invaders were more distantly related to natives at very small scales (<4 m^2) but clustered with them at moderate to large scales (64 m^2 and 35 km^2). Further strengthening this observation of the domination of underdispersion at moderate to large scales, Strecker and Olden (2014) showed that non-native fish species are clustered with natives at watershed scales, which presumably contain a number of different types of rivers, streams, and wetlands, each supplying environmental filters to the regional biota. Among non-native plant species inhabiting a national park outside of Sydney, Australia, Cadotte, Hamilton, et al. (2009) showed that occupancy patterns (number of grid cells occupied) were phylogenetically clustered at the continental scale, but there was no signal at more local scales within the park. The reason for this was that shared climatic tolerances informed where related species could occur, but responses to small-scale processes within the park (disturbance, stochasticity, dispersal, competition, etc.) generally lacked phylogenetic signal among species.

While there is a logical set of assumptions that seem to support the switch from overdispersion at small scales to clustering at large scales, tests of this pattern identify important exceptions. These include clustering at small scales and/or overdispersion at large scales, which can each offer insights into the role that different structuring mechanisms play in community assembly. At extremely large spatial scales (i.e., continental scales), bird assemblages may be overdispersed in some regions and clustered in others (Barnagaud et al.

2014). This change in phylogenetic patterns appears to reflect the distribution of the traits driving ecological patterns. Carnivorous and migratory birds tend to be phylogenetically clustered, whereas frugivores and nectavores are phylogenetically overdispersed (Barnagaud et al. 2014). Similarly, vertebrates in Australia also show more idiosyncratic phylogenetic patterns at large scales (Lanier et al. 2013). These examples highlight that switching between overdispersion to clustering can also reveal the evolution of different strategies and the geographical locations where different strategies have been successful.

There are a number of examples where local communities are clustered and nested within larger scales that are themselves clustered. From chapter 3 we have seen that mechanisms at the local scale can generate clustered local communities (e.g., Helmus et al. 2010, Oke et al. 2014), but does this translate to larger spatial scales? Small- and large-scale clustering seems to be the dominant pattern observed in disturbance-prone coastal sand dunes (Brunbjerg et al. 2012, Brunbjerg et al. 2014). Here the particular system includes strong selection for species with key disturbance tolerant traits at small scales. At large scales, these sand dunes are in a mosaic with other habitat types and there are sand dune specialists that come from relatively few lineages. In these types of disturbance-dominated systems, local communities may be phylogenetically clustered because competition among species neither has the time nor stability to realize competitive exclusion; thus, groups of closely related species adapted to local disturbances will be able to persist in local communities, resulting in clustered assemblages (fig. 7.12). At larger scales, it could be that the same selection processes simply scale up, or there may be a more general trend for weaker competition at larger spatial scales. More research testing scale-dependent mechanisms on clustered communities is required.

Interestingly, mobile species may be clustered for entirely different reasons, and may show clustering at multiple scales as well. Harmon-Threatt and Ackerly (2013) show that bee communities are phylogenetically clustered at multiple spatial scales because their mobility allows them to reduce competition while maximizing the fit between traits and local resources. Thus similar species appear to coexist spatially, but might be overdispersed temporally.

One interesting and additional contradictory pattern has begun to emerge, and requires more comparative studies. Trees in tropical forests often appear to be clustered or random at small scales (Webb 2000) and potentially overdispersed at larger scales. Local clustering could occur because tropical trees are extremely specialized, with fine-scale partitioning of niche space (Futuyma 2010), and respond to small-scale environmental variation (see also chapter 3 for further discussion on clustering at small scales). The large-scale overdispersion may reflect past speciation events or convergent evolution among disparate lineages. Random patterns, however, might suggest either a mix of structuring processes or neutral dynamics (Hubbell 2001).

7.2.1.1. A Guide for Testing Scale Dependency in Phylogenetic Patterns

It is technically straightforward to explore patterns at multiple scales (e.g., following the methods outlined in chapters 3 and 4). Essentially, we look for phylogenetic patterns; for example, we look for over- and underdispersion at the smallest scale, then combine plot data together into cumulative sets representing some larger scale, and run the analysis again. The key is how to construct the species pool for the randomizations. The pools could

A Nested design   **B** Random design

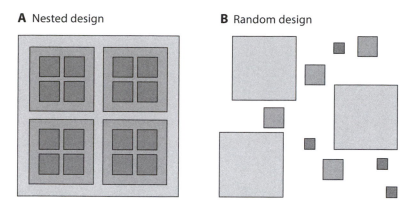

Figure 7.13. The two primary designs used to examine the relationship between scale and diversity.

be from the largest scale analyzed, or from some regional species list (see also chapter 4 on selecting the correct species pool).

Many of the studies outlined above implicitly make differing assumptions about the dominant mechanisms being tested, and this is manifest in how they scale up their observations and construct species pools. One obvious question is whether independent plots should differ in size, or whether they should be nested (fig. 7.13).

Plots of differing sizes are useful for constructing diversity-area relationships (see section 7.2.2) and may be powerful for detecting scale dependency in phylogenetic patterns. Independent and randomly distributed plots (Fig. 7.13B) of different sizes ensure that smaller-scale patterns are not influencing larger-scale ones (e.g., nonindependence). Nested designs (Fig. 7.13A) have a pseudoreplication problem (Hurlbert 1984) and suffer from nonindependent spatial samples of unequal sample size. However, the nested designs are easier to sample (larger scales are just combinations of smaller plots), and it is clearer what the species pool ought to be (the cumulative species pool). Further, the independent and random design imposes a limit on the largest size patch that can be realistically sampled, whereas with the nested design, one could conceivably combine plots all the way up to the global scale. The nested design is most frequently employed.

Even accepting problems regarding nested designs, there remain substantial methodological issues that can bias results toward observing particular patterns. Diane Srivastava (1999) outlined how underappreciated methodological decisions influenced observations of the relationship between local and regional richness, and here we take her lead to outline how similar decisions may influence observations of phylogenetic patterns and the resulting inferences (fig. 7.14). The most critical methodological decision is whether studies include a single habitat type, or whether larger scales combine multiple habitat types. The next decision is whether plots are combined within a single region or across multiple regions. If multiple habitats are selected (e.g., lakes, ponds, rivers, and wetlands), then environmental heterogeneity is likely to be the dominant mechanism shaping species assembly, and observations of clustering likely reflect the fact that habitat preference is a deeply conserved trait. However, this design confounds analyses of scale dependency, because clustering could also be observed at very small scales if substantial environmental gradients exist, and likely

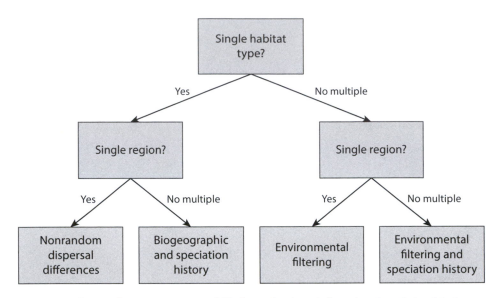

Figure 7.14. Geographic comparisons and likely mechanisms influencing the relationship between spatial scale and diversity.

obscures how phylogenetic patterns change with increasing scale. If a single habitat type is selected, then phylogenetic patterns are more likely driven by dispersal differences and biotic interactions, such as the presence/absence of key species (e.g., mutualists, prey, competitors, etc.). Scaling up across multiple regions, even within a single habitat type, introduces the complication that biogeographic and speciation histories can differ in idiosyncratic ways.

Before evaluating the relationship between spatial scale and community structure, it is perhaps informative to first consider how phylogenetic diversity scales with area.

7.2.2. Phylogenetic Diversity-Area Relationships (PDAR)

How phylogenetic diversity is related to geographical area is currently a nascent area of research, but shows promise as a way to uncover potential mechanisms influencing phylogenetic patterns. The observation that the numbers of species increase with area sampled is an old one (Arrhenius 1921), and ecologists continue to debate its meaning—questioning whether it is simply a statistical expectation or if it reflects meaningful ecology (Gleason 1922, Connor and McCoy 1979, Crawley and Harral 2001, Drakare et al. 2006, Whittaker and Triantis 2012). The core of the species-area relationship (SAR) is a relatively simple power function:

$$S = cA^z \tag{7.9}$$

Here S is the number of species in a habitat of area A, c is a constant equaling the number of species in the smallest unit (e.g., 1 km² if kilometers is the unit used in the analysis), and z is the slope in log-log space. The SAR can be represented by a simple linear model:

$$\log(S) = \log(c) + z \cdot \log(A) \tag{7.10}$$

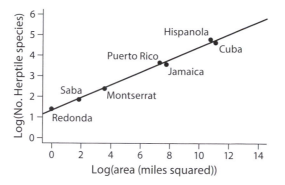

Figure 7.15. The relationship between species richness and area for Caribbean islands. From Macarthur and Wilson (1967).

The result is a linear relationship between area and richness (fig. 7.15).

The slope z is thought to reflect important aspects of organismal biology and ecology. Specifically, body size, mobility, and life history, as well as habitat connectedness can all increase or decrease how well organisms move between patches (Holt et al. 1999, Drakare et al. 2006). Thus, the SAR for highly mobile organisms living within well-connected patches should be flatter, with a lower z and higher c.

How does this long history of SARs translate into phylogenetic diversity-area relationships (PDAR)? The answer is not straightforward. There are three main challenges: (1) which metric to use, (2) how to model the PDAR relationship, and (3) how to infer mechanisms from the shape of the PDAR.

7.2.2.1. Metric Choice

To examine PDAR we first must select an appropriate metric to use. In chapter 3 we reviewed two-dozen metrics that have their own behaviors and are variously correlated with richness. The fact they behave differently means the PDARs will be quite different from one another. PD is an obvious candidate to plot against area, as it is correlated with richness and we should expect it to increase monotonically with area (Morlon et al. 2011, Rodrigues et al. 2011). Researchers have asked how this type of PDAR differs than that expected from the SAR (Morlon et al. 2011, Rodrigues et al. 2011, Mazel et al. 2014). It is clear that this PDAR is not linearly related to area, but instead is a saturating curve (fig. 7.16) based on a random expectation of community PD (Morlon et al. 2011). An alternative measure that is uncorrelated with richness—MPD—appears virtually unchanged across area (fig. 7.16). We can examine these metrics using the Jasper Ridge species list and phylogeny, and functions in *picante*:

Let us specify the expected number of species (S) across an arbitrary gradient in spatial scale (A) using the SAR, giving reasonable values to z (z is often estimated at between 0.25 and 0.3) and c:

```
z<-0.25
A<-exp(seq(2:10))
```

```
c<-3
S<-round(c*(A^z))
```

Here the `round()` function rounds the estimated values to whole numbers, just to keep things simple. Now let us calculate PD and MPD for randomly assembled communities for each habitat size. First we create a data matrix filled with zeros to store our values for the simulated Jasper communities:

```
tmp<-matrix(0,100,ncol(j.com))
colnames(tmp)<-colnames(j.com)
```

And we specify the number of random communities at each habitat size class:

```
nrand<-100
```

Then we establish the containers to hold our data:

```
PDAR<-NULL
mPDAR<-NULL
```

Now we can run a nested "for" loop that operates across the different area size classes and randomizes `nrand` assemblages within each class. In this loop, the `sample()` function is randomly placing 1's in the matrix according to the estimated species richness (S):

```
for (i in 1:length(A)){
    for (j in 1:nrand){
        tmp[j,sample(1:ncol(tmp),S[i])]<-1
    }
    PDAR<-c(PDAR,pd(tmp,j.tree)$PD)
    mPDAR<-c(mPDAR,mpd(tmp,cophenetic(j.tree)))
}
```

We also need to create a vector of area values (A), each replicated by the number of randomizations (`nrand`) for plotting:

```
A.out<-rep(A,each=nrand)
```

Instead of a scatterplot, we will use a lowess-smoothed line to plot the data so as to make the general behaviors more apparent (fig. 7.16).

Mazel et al. (2014) took a different approach; they compared the behavior of several versions of the generalized Hill number (see chapter 3), where q = 0 is equivalent to a standardized Faith's PD, PD_{SND}. This is PD/T, where T is the root to tip distances; q = 1 is a standardized entropic measure; and q = 2 is a standardized Rao's diversity measure (mentioned previously in chapter 3). The use of Hill numbers here is important, because Hill numbers represent the "effective number of species" and can be directly compared to richness (Chao et al. 2010). Figure 7.17 shows the results of one of the analyses from Mazel and colleagues (Mazel et al. 2014); each of the three Hill number metrics is plotted as PDAR minus SAR, so that a positive number means that the PDAR is larger than SAR and a value of zero means they are equivalent. Their results show that PDARs generally rapidly increase and then saturate, much like in our figure 7.16.

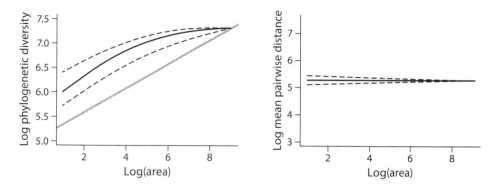

Figure 7.16. Diversity-area relationships from simulated data using the Jasper Ridge phylogeny. The *black lines* indicate the values of the phylogenetic measure, the *dashed lines* show the maximum and minimum values, and the *solid gray line* shows how richness changes with area.

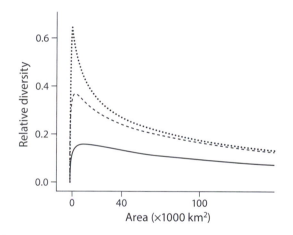

Figure 7.17. The behavior of three phylogenetic measures across area. In this figure the *solid line* shows Faith's PD ($q = 0$); the *dashed line* shows Allen's diversity ($q = 1$); and the *dotted line* shows Rao's diversity ($q = 2$). The q values refer to the exponent from the Hill numbers equation. Adapted from Mazel et al. (2014).

7.2.2.2. Fitting the Curve

Unlike SARs, there are no underlying mechanistic explanations for PDARs and it is not clear how to fit statistical models to them. Morlon et al. (2011) fit a power function much like the traditional SAR, but with a slight difference:

$$PD = T_0 c^{z'} A^{z \cdot z'} \tag{7.11}$$

where T_0 is the most recent common ancestor distance in the phylogeny and z' is the exponent for the PD-area relationship. Mazel et al. (2014) argue that we don't have sufficient information to assign a model *a priori*, and instead advocate comparing multiple PDAR models, listed in table 7.2. In their analysis, they found that different models best fitted PDARs in different habitat types.

TABLE 7.2.
Examples of species-area functions, including the formula, curve shape, and whether the curve reaches an asymptote.

Function	Formula*	Curve shape	Asymptote?
Power	$S = cA^z$	convex	No
Extended power	$S = cA^{zA \cdot d}$	convex or sigmoid	No
Exponential	$S = c + z \cdot \log A$	convex	No
Monod	$S = d/(1 + cA^{-1})$	convex	No
Logistic	$S = c/(f + A^{-z})$	sigmoid	Yes
Weibull	$S = d(1 - \exp(-cA^z))$	sigmoid	Yes
Asymptotic	$S = d - cz^A$	convex	Yes
Rational	$S = (c + zA)/(1 + dA)$	convex	Yes

*Terms: S = species richness or other diversity measure (e.g., PD); A = area; c, d, f, z = fitted parameters.
Source: Mazel et al. 2014.

At this point in time there has not been a thorough comparison of the scale dependency or independency of different metrics. Further, there is no current theory that identifies a particular model to explain PDARs (Wang et al. 2013). We suggest there is much scope to explore these questions in the future.

7.2.2.3. Inference from Shape of PDARs

Despite the current lack of theoretical underpinning, a few studies have attempted to use simulations and null models to reject particular mechanisms that may shape PDARs (Helmus and Ives 2012, Wang et al. 2013). Wang et al. (2013) used point-process models to generate null models that were completely random or included either dispersal limitation, habitat filtering, or both. Data from two forests supported the dispersal limitation model as the most likely explanation of observed PDARs. The Wang et al. (2013) approach is an intriguing one, and no doubt more work will be done on this topic.

7.3. CONCLUSION

Spatial ecophylogenetic patterns can reveal how different processes shape ecological communities. Metacommunity ecology (Leibold et al. 2004, Holyoak et al. 2005) reinvigorated the search for general mechanisms that create diversity and structural differences among communities, and by using phylogenetic patterns we can understand how shared traits and evolution inform community differences (Dray and Legendre 2008). One set of metacommunity methods employs distance matrices (e.g., compositional dissimilarity, spatial distances, etc.; see Cottenie 2005), integrating a phylogenetic distance matrix thus seems natural and straightforward (Pavoine et al. 2011). However, as we have discussed in this chapter, there are a number of methodological decisions that are required, and interpretation of patterns can be less than straightforward.

More importantly though, there are more basic questions about how phylogenetic patterns change across spatial scales that have been under-researched thus far. For example, how does evolutionary history or biogeography affect the accumulation of phylogenetic diversity as we increase the scale of our observation? Ecologically, we assume that different mechanisms dominate at different scales, but this has rarely been tested, and ways in which these mechanisms are reflected in real landscapes can be quite varied. Do we expect the same type of mechanism switch in a homogenous landscape as for one with a high degree of heterogeneity? For that matter, at what scale do we see this switch?

Community ecology and phylogenetics have merged nicely over the past decade. However, we believe that there are still many opportunities for merging spatial ecology with phylogenies.

CHAPTER 8

Speciation, Extinction, and the Distribution of Phylogenetic Diversity

Much history of the world has been lost.
How magnificent and confusing what's left is.
—James Dye

The diversity of life is unevenly distributed across the globe. Species richness tends to be higher at lower latitudes and elevations, although the underlying explanations continue to be debated (Willig et al. 2003). The distribution of life forms also varies across space. For example, Bergmann's rules suggests that mean body sizes increase in colder climates, and Foster's rule (also referred to as the island rule) suggests that on islands small species evolve to become bigger, while large species evolve to become smaller. Equally, the distribution of evolutionary history shows large spatial variation, reflecting the histories of speciation, extinction, and dispersal. This spatial variation can thus provide insights into the underlying mechanisms driving biodiversity gradients. In addition, hotspots of evolutionary history might represent targets for conservation aimed at preserving the tree of life.

8.1. CONSERVATION OF THE TREE OF LIFE

There has been a growing call to consider evolutionary history in conservation planning; see Purvis et al. (2005) and for a more recent overview see Rolland et al. (2011). For example, if phylogenetic diversity captures feature diversity that is useful for maintaining ecosystem processes (see chapter 9), or if it provides insurance against an unknown future (Faith et al. 2010) then conservation efforts should not just focus on preserving species; they should also focus on the branches of the evolutionary tree from which they descend (Mace et al. 2003). Indeed, some of the earliest phylogenetic indices, including Vane-Wright's measure of taxonomic distinctiveness (Vane-Wright et al. 1991) and Faith's PD (Faith 1992a), were original presented within a conservation framework. Large, global phylogenies allow us to map the distribution of phylogenetic diversity, and develop a conservation strategy to maximize coverage of the tree of life.

   Although many schemes have been developed for incorporating phylogenetic information into conservation planning, ranging from regional to global scales (see chapter 9), there has been a conspicuous absence of uptake by conservation practitioners. In part, this disconnect reflects a lack of communication between researchers and conservation practitioners, and a jumble of terms and metrics that have added confusion to the field (Winter et al. 2013). However, additional criticism stems from the fact that PD and species richness are closely correlated, which would seem to diminish the value of PD as a unique conservation measure (Rodrigues et al. 2005, Rodrigues et al. 2011). We can easily demonstrate this correlation using the distribution of mammals in Australia as an example. Here, using a subset of the global mammal data published in Davies and Buckley (2012), we can plot the distribution of species richness from a presence-absence matrix of species by cells (data is available at: http://press.princeton.edu/titles/10775.html):

```
mat<-read.table("australia.txt", header=T)
```

We can calculate species richness per cell simply by summing the rows and making sure that cell identities (provided in the row names of the matrix) are retained:

```
rich<-rowSums(mat)
rich<-cbind(as.numeric(as.vector(rownames(mat))), rich)
colnames(rich)<-c("id", "richness")
```

In this data set, cell identities are referenced using index numbers, so we need to provide spatial (latitude and longitude) coordinates to be able to plot the data. In this example this information is in a separate translation table that we can merge with the species matrix on cell IDs:

```
latlong<-read.table("cell translation.txt", header=T)
richness<-merge(rich,latlong, by = "id")
```

We now have a table with species richness and spatial coordinates per cell. Note that the spatial coordinates represent cell center points within a geographic coordinate system (latitude and longitude in decimal degrees). These data can be exported to a GIS program, such as ArcMap or freeware equivalents (e.g., GRASS: http://grass.osgeo.org/ or QGIS: http://www.qgis.org). Recently, several libraries have been developed in R for handling and displaying spatial data (e.g. *sp, raster, rasterVis, maptools, rgeos*); many of these tools and the analysis of spatial data have been reviewed elsewhere (e.g., Bivand et al. 2008). For illustration we can generate a simple plot using the *raster* library.

First, we need to define the bounds of our coordinate space—that is, the maximum and minimum latitudes and longitudes:

```
xmax <-max(richness$longitude)
xmin <-min(richness$longitude)
ymax <-max(richness$latitude)
ymin <-min(richness$latitude)
```

Next we generate a rectangular array with rows representing latitude and columns longitude and populate it with our richness values using a simple "for" loop:

```
aus.array<-array(NA, c(length(unique(richness$latitude)),
        length(unique(richness$longitude))))
colnames(aus.array)<-sort(unique(richness$longitude))
rownames(aus.array)<-sort(unique(richness$latitude),
        decreasing = T)

for (k in 1: length(richness[,1])){
aus.array[which(rownames(aus.array)==richness[k,4]),
        which(colnames(aus.array)==richness[k,3])]
        <-richness[k,2]
}
```

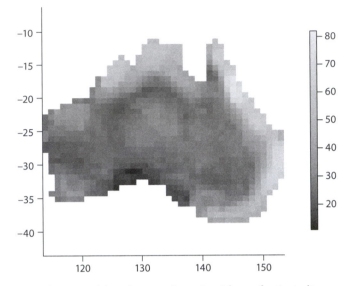

Figure 8.1. Map of mammal species richness for Australia.

The dimensions of the array match the number of unique latitude and longitude values in our data. Note that we have been careful to name the rows (latitude) and columns (longitude) in our array because this enables us to fill it by matching the longitudes and latitudes from our table of species richness (columns three and four, respectively) using the `which()` function.

Finally, we convert our array into a *raster* object, specifying the bounds to our coordinate space, which can be displayed using the `plot(raster)` function:

```
aus.raster<-raster(aus.array, xmn = xmin, xmx = xmax, ymn =
     ymin, ymx = ymax)
plot(aus.raster))
```

In the plotted data (fig. 8.1) we can see clearly the band of high species richness (light shading) matching closely to the distribution of the tropical and temperate forest biomes, with low species richness characterizing the dry, desert landscape of the interior—Australia's famed "outback." While this steep diversity gradient does not follow the global latitudinal gradient in species richness, it matches the pattern of high diversity in warm and wet (tropical) climates.

Now let us compare patterns of phylogenetic diversity (Faith's PD), which is a derived product of the species assemblage. First, we need to calculate PD for each cell. Let's generate a three-column matrix with `nrows` equal to the number of cells to store our results, and populate the latitude and longitude columns:

```
PD<-matrix(0, nrow(mat),3)
row.names(PD)<-row.names(mat)
colnames(PD)<-c("PD","Long","Lat")
```

Now we add latitudes and longitudes:

```
match1<-match(row.names(PD), latlong$id)
PD[,2]<-latlong[match1,"longitude"]
PD[,3]<-latlong[match1,"latitude"]
```

Next we need to import the mammal tree for Australian taxa:

```
phy<-read.tree("australia.tree")
```

Now we must loop through each cell in the matrix, extract the list of species that have overlapping distributions, and calculate the sum of the branch lengths for the subtree that links them together:

```
for (i in 1:nrow(mat)) {
    samp = mat[i,]
    sp.drop= names(samp)[which(samp==0)]
    if(sum(samp)>1){ #check at least 2 species
    sub.tree<-drop.tip(phy,sp.drop)
    PD[i,1] = sum(sub.tree$edge.length)
        } #end check 2 species
} #end loop cells
```

Note that within this loop we have a nested `if` statement that checks whether there are at least two species in each cell; this is because a phylogeny of one species is not meaningful, and PD is calculated as the sum of the branch lengths excluding the root branch, and thus a single species cell has PD = 0. We can modify this preference; one approach would be to include a dummy species that occurs in all cells and affix it to the root of the tree with a zero branch length.

Finally, before plotting, we convert our matrix to a data frame, and label columns appropriately:

```
PD<-as.data.frame(PD)
PD<-cbind(row.names(PD), PD)
colnames(PD)<-c("id", "PD", "longitude", "latitude")
```

We can fill our array as we did for species richness:

```
for (k in 1: length(PD[,1])){
aus.array[which(rownames(aus.array)==PD[k,4]), which
    (colnames(aus.array)==PD[k,3])]<-PD[k,2]
}

PD.aus.raster<-raster(aus.array, xmn = xmin, xmx = xmax, ymn
    = ymin, ymx = ymax)
plot(PD.aus.raster)
```

Most obviously, we can see that the distribution of PD is almost indistinguishable from the distribution of species richness (compare figs. 8.1 and 8.2). This strong covariation can also be simply illustrated by correlating one against the other:

```
data<-merge(PD, richness, by = "id")
plot(data$richness, data$PD)
cor(data$richness, data$PD)
```

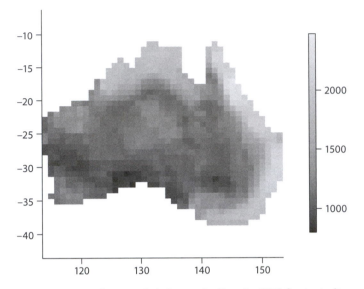

Figure 8.2. Map of mammal phylogenetic diversity (PD) for Australia.

Phylogenetic diversity (PD) shows a strong, but saturating relationship with species richness (we would expect this from reading chapter 7), and a correlation coefficient of $r =$ 0.98 (fig. 8.3). It seems we have not gained much from the addition of phylogeny into our analysis; selecting areas for conservation based on richness or PD would lead to the identification of similar priority areas. As both PD and species richness represent summations, this result is perhaps not unexpected (Davies and Cadotte 2011, Rodrigues et al. 2011). However, conservation decision making is typically focused on marginal gains, and plays out in a cost-benefit scenario. If every species costs the same amount of money to conserve (an admittedly unlikely scenario) we might be more interested in selecting areas that capture higher phylogenetic diversity than predicted from species richness. One approach would be to regress PD against richness and look at the residuals; although the saturating relationship between richness and PD adds complexity, and in chapter 4 we discuss reasons why this approach might not be most appropriate. Davies et al. (2008) fitted a spline (a piecewise polynomial function) to model the relationship, but the choice of appropriate "smoothing" is somewhat arbitrary (see chapter 7 for further discussion). An alternative is to compare observed PD values against expectations from resampling the same number of species at random across the phylogeny (see chapter 4 on null models). As we have already seen, a standard effect size (SES) for PD can then be calculated straightforwardly. We could write a short loop to calculate standard effect sizes ourselves using the `sample()` function; however, as we have seen in chapter 3, the `ses.pd()` function in *picante* is already available to us:

```
PD.ses<-ses.pd(mat, phy, null.model = c("taxa.labels"),runs
    = 99, include.root=FALSE)
```

Here, for computational ease, we have specified 99 runs (randomization) because we are working on a large number of species and grid cells, but we might want to increase this by another order of magnitude for publication (for further discussion see chapter 4). The

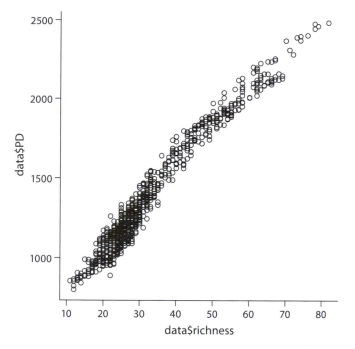

Figure 8.3. Scatterplot of species richness against phylogenetic diversity (PD) for Australian mammals sampled on a geographic grid.

SES values can plotted as described above, first linking the cell IDs back to their spatial coordinates. We are interested in the sixth column in the results from `ses.pd()`, headed `pd.obs.z`; these z-values represent the SES for PD:

```
ses.data<-cbind(as.numeric(rownames(PD.ses)), PD.ses$pd.
    obs.z)
colnames(ses.data)<-c("id", "SES")
SES<-merge(ses.data,latlong, by = "id")
```

We can plot our SES values exactly as we did for PD:

```
for (k in 1: length(SES[,1])){
aus.array[which(rownames(aus.array)==SES[k,4]),
    which(colnames(aus.array)==SES[k,3])]<-SES[k,2]
}

SES.aus.raster<-raster(aus.array, xmn = xmin, xmx = xmax,
    ymn = ymin, ymx = ymax)
plot(SES.aus.raster)
```

Diversity patterns based on the SES values for PD (fig. 8.4) now look very different from those for PD or species richness. High SES values represent cells where the mammal assemblage captures more PD than expected by chance, as estimated by sampling the same number of species at random from the phylogeny. These results show that regions with high richness and PD also represent regions where mammal assemblages have a high

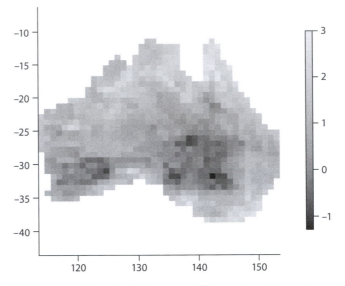

Figure 8.4. Map of phylogenetic diversity (PD), corrected for species richness (`ses.PD`), for Australian mammals.

evolutionary distinctness. In contrast, low diversity regions capture less evolutionary history than expected, suggesting the mammal assemblages found there are comprised of a few closely related species. It is, however, notable that most cells have positive SES values, as we can see from the histogram of their frequency distribution (fig. 8.5), suggesting a continental trend toward more evolutionarily dispersed assemblages:

```
hist(PD.ses$pd.obs.z)
```

At local scales we might predict patterns of phylogenetic overdispersion due to competitive exclusion among close relatives (chapter 7). However, at regional scales such an explanation seems unlikely, and we might have predicted there to be a greater tendency toward underdispersion—for example, as local radiations would generate geographic clusters of closely related species (chapter 7). Our result might, therefore, seem surprising. One possible explanation relates back to the underlying phylogeny of Australian mammals, which includes two monotremes (the duckbilled platypus, *Ornithorhynchus anatinus,* and the short-beaked echidna, *Tachyglossus aculeatus*) subtending from a very long evolutionary branch. If both species are relatively widespread, such that the majority of cells have one or the other or both, then observed phylogenetic diversity will be relatively high; however, the probability of selecting a monotreme in a particular randomization is relatively low (approximates $1/119*n$,[1] where there are 2 monotremes out of 238 tips in the phylogeny, and n is the number of species in the cell assemblage). Thus any cell with a monotreme will

---

1. We can calculate the exact probability using the following script, kindly provided by Frédéric Boivin:

```
Temp = marsupial/(marsupial+monotreme)
for (i in 1:(n-1)){
Temp=Temp*(marsupial-i)/((marsupial+monotreme)-i)
}
probability <- 1 - Temp
```

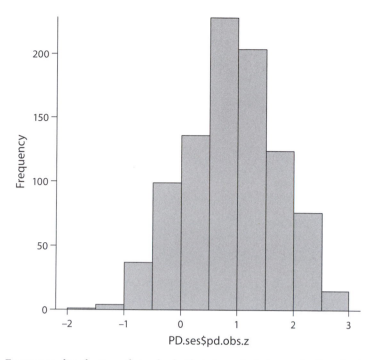

Figure 8.5. Frequency distribution of standard effect sizes of phylogenetic diversity (`ses.PD`) for Australian mammals.

tend to have high SES. It is notable that the echidna has a wide range and is found across most of Australia. Our results might therefore reveal more about our choice of null model (randomizing tips), which ignores the relative frequencies of species, than it does about the underlying phylogenetic structure of species assemblages (see chapter 4 for further discussion on null model choice).

We can further explore patterns by repeating the analysis excluding monotremes:

```
phy2<-drop.tip(phy, c("Tachyglossus_aculeatus",
    "Ornithorhynchus_anatinus"))
mat2<-mat [, which(!colnames(mat) %in% c("Tachyglossus_
    aculeatus", "Ornithorhynchus_anatinus"))]
```

While we see geographic gradients are broadly similar (compare figs. 8.4 and 8.6), species-rich areas capture relatively greater phylogenetic diversity than species-poor areas, but the mode of the SES has shifted from positive to negative values (compare figs. 8.5 and 8.6). This trend is more consistent with expectations based on historical biogeography and phylogenetic niche conservatism (Wiens and Donoghue 2004).

8.1.1. Complementarity

Hotspot approaches have attracted much attention in global conservation biology (Myers et al. 2000). Conservation International's 35 Biodiversity Hotspots are defined largely on endemic richness, such that the overlap in species composition is relatively small

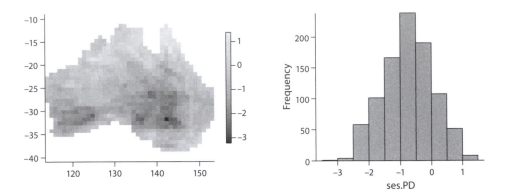

Figure 8.6. Map of phylogenetic diversity corrected for species richness (`ses.PD`) for Australian mammals excluding monotremes (*left*), and matching frequency distribution of SES (standard effect size) values (*right*).

between them. However, when prioritizing sites regionally, widespread species are likely to occupy several sites. Therefore, when selecting a subset of sites from many possible sites, the order in which they are selected matters. Simply ranking sites by richness and prioritizing the highest ranked sites can be suboptimal. For example, it is possible that sites ranked two and three might, in combination, capture greater species richness than the top two most species-rich sites, especially if there is high similarity in species composition between them (Kirkpatrick 1983, Margules et al. 1988). Site selection to maximize the total number of represented species is based on the concept of complementarity, where each site selected adds maximally to (i.e., maximally complements) the total species set (Vane-Wright et al. 1991, Pressey et al. 1993). Selecting the optimal reserve set is a notoriously difficult problem when the number of possible sites is large. There are now a number of custom-made software programs, such as MARXAN, C-PLAN, ZONA-TION, and so forth, that consider costs as well as benefits in reserve selection (Sarkar et al. 2006, Pressey et al. 1997, Margules and Pressey 2000). However, the principle of complementarity can be easily demonstrated using the "greedy" algorithm, which chooses sites myopically so as to maximize the immediate gain in diversity with the addition of a site given an existing selection—rather than searching for the global optimum set of *n* sites.

Annotated code is provided in the online supplement, because it is somewhat unwieldy; however, it follows a simple heuristic:

Step one: select the site that captures the most diversity.
Step two: select a new site that adds the most additional diversity to the selected set.
Step three: repeat step two until either all diversity is captured, or until some predefined maximum number of sites has been selected.

We can show how cell selection based on optimizing PD differs from that when the criterion is species richness by plotting the ranks of the first 20 selected cells under each optimization criterion (fig. 8.7). Here the cells are ordered according to their rank based on PD, and on their matching rank when cell selection is optimized on gain in richness. Gaps represent cells selected in the top 20 cell-set for PD that were not in the richness set. We can

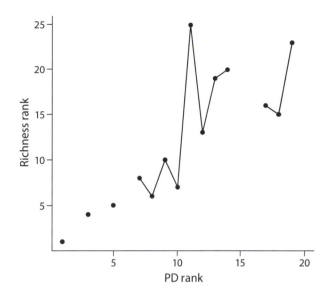

Figure 8.7. The top 20 ranked cells selected using complementarity to maximize species richness, plotted against their matching ranks when selected to maximize phylogenetic diversity (PD). Cells are ordered by their richness ranks (x-axis).

thus see clearly that while overall performance might be similar for both criteria, in practice the choice of cells can be very different:

```
ranks <-match(results.SR$ID, results.PD$ID)
plot(ranks[1:20], pch = 16, type = "l")
points(ranks[1:20], col = "red", pch = 16)
```

8.2. MACROEVOLUTION: DIVERSIFICATION

While the use of phylogenetic methods in community ecology is relatively recent, evolutionary history has long been argued to be important in explaining large-scale gradients in species richness (see review in Mittelbach et al. 2007). Regional variation in species richness is a product of three fundamental processes: speciation, extinction, and migration (Godfray and Lawton 2001). At least the first two, and arguably the last (e.g., see Davies and Buckley 2012) might leave an imprint on the phylogeny of extant taxa. The latitudinal gradient in species richness is one of the most widely recognized spatial biodiversity patterns, and it is replicated across taxonomic groups (Hillebrand 2004) as well as over paleontological time (Willig et al. 2003, but see Mannion et al. 2014). The botanist G. Ledyard Stebbins suggested the high species richness in the tropics might be explained by one of two processes. The first was that the tropics might be an evolutionary cradle, where rapid speciation leads to high species richness; the second was that the tropics might be a "museum" of diversity, where extinction rates have been slow, and species richness has had a long time to accumulate (Stebbins 1974). The availability of large, near complete, phylogenetic trees for species-rich groups, such as mammals (Davies and Buckley 2012), birds (Weir and Schluter 2007, Jetz et al. 2012), amphibians (Pyron and Wiens 2013), and some plant clades (Davies et al. 2004), has allowed some of the first direct tests of these competing hypotheses.

If hotspots of high species richness represent museums of diversity, we might expect species ages to be older, on average, where diversity is high. We can extract approximations of species' ages from phylogeny by taking their pendant edge lengths. However, true species ages may be biased by recent extinctions; for example, the extinction of one species from a sister pair will suggest an older age for surviving species. We might nonetheless expect that the distribution of pendant edges would be skewed shorter if speciation has been rapid. We can extract pendant edges directly from the phylogeny as follows, again using Australian mammals as an example:

```
sp<-NULL
edge<-NULL
for (x in 1: length(phy$tip.label)){
sp<-c(sp, phy$tip.label[x])
edge<-c(edge, phy$edge.length[phy$edge[,2]==x])
}
species.ages<-cbind.data.frame(sp, edge)
```

Now let us plot the median age (fig. 8.8):

```
ages<-matrix(0, nrow(mat),3)
row.names(ages)<-row.names(mat)
colnames(ages)<-c("ages","Long","Lat")

#add lat and lon
match1<-match(row.names(ages), latlong$id)
ages[,2]<-latlong[match1,"longitude"]
ages[,3]<-latlong[match1,"latitude"]

for (i in 1:nrow(mat)) {
    samp = mat[i,]
    sp.in.cell<-names(samp)[which(samp==1)]
    cell.ages<-species.ages[species.ages$sp %in% sp.in.
        cell,]
    ages[i,1] = median(cell.ages$edge)
} #end loop cells

ages<-as.data.frame(ages)

ages<-cbind(row.names(ages), ages)
colnames(ages)<-c("id", "age", "longitude", "latitude")

for (k in 1: length(ages[,1])){
aus.array[which(rownames(aus.array)==ages[k,4]), which
        (colnames(aus.array)==ages[k,3])]<-ages[k,2]
}

ages.aus.raster<-raster(aus.array, xmn = xmin, xmx = xmax,
        ymn = ymin, ymx = ymax)
plot(ages.aus.raster)
```

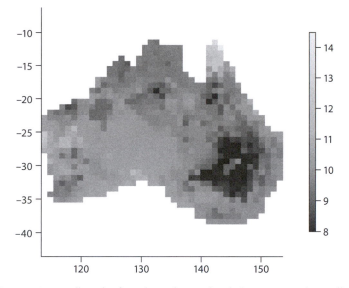

Figure 8.8. Median species age (length of pendant edge on the phylogenetic tree) in millions of years for Australian mammals.

We can see (fig. 8.8) that only the Cape York Peninsula in Queensland might be considered a museum of diversity, as this region is both species rich and contains taxa that are on average older than taxa found elsewhere. While species ages might capture recent diversification, the history of speciation and extinction reaches deeper into the phylogeny. Jetz et al. (2012) showed that per lineage diversification rates, which integrate the birth and death of lineages from the root to each tip separately, are equivalent to 1/ED, where ED (evolutionary distinctiveness; see chapter 3) is calculated using the equal splits measure of Redding and Mooers (2006). This measure can be extracted as follows (see also chapter 9):

```
ED<-evol.distinct(phy, type = c("equal.splits"))
```

We can see clearly (fig. 8.9) that our original conclusion is supported, and that hotspots of richness do not represent cradles of diversity, but are more consistent with evolutionary museums; the fastest per lineage diversification rates are found where species diversity is generally low. We should be careful here in that metrics of ED assume that all diversity is sampled; the example of Australia was chosen because most mammal diversity within the continent is a product of *in situ* radiation. Nonetheless, it is important that results of these types of analyses are interpreted with care when complete phylogenies are not used. In our example, Australian mammals include monotremes, marsupials, and a few placental clades (i.e., bats and rodents) and thus do not comprise a monophyletic clade. In addition, extant mammal diversity in Australia represents the remnants of a much more diverse fauna. Before we rush to accept evolutionary explanations, we might first want to consider historical and environmental interpretations. For example, if the tropical biome was once more extensive, but has declined recently, we might expect current tropical regions to have higher species richness if environmental niches have been conserved over evolutionary time (Wiens and Donoghue 2004). In addition, we might also expect fewer older species within younger biomes if the age of these biomes is younger than the typical age of a species.

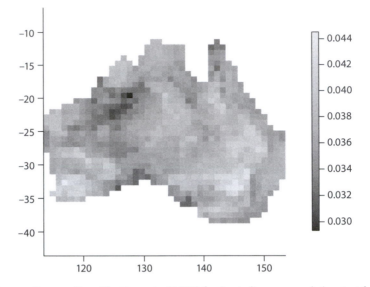

Figure 8.9. Mean per lineage diversification rates (1/ED) for Australian mammals (see text for details).

We note that it is possible to plot the distribution of any metric that can be calculated at the species level; for example, evolutionary isolation, threat, or both (see e.g., the EDGE metric of Isaac et al. (2007)). It is also possible to weight species by their abundances or economic importance, and this is done by multiplying the species presence-absence matrix by the relevant scalar.

8.2.1. Diversification through Time

In the preceding sections of this chapter, we explored the variation in diversification across space; however, accurately dated phylogenies also allow us to explore rates through time. As in our analyses above, a first step is to visualize the data, and here we can use lineage through time plots (LLTs) to show the rate of lineage accumulation. Keep in mind that we are only considering extant diversity, and that at any time in the past the true number of species was likely much greater than suggested by the number if lineages ancestral to extant species. Indeed, under some evolutionary models, standing diversity may have been relatively constant through time.

Here is the LTT plot for Australian mammals on a semi-log plot (fig. 8.10):

```
ltt.plot(phy, log = "y")
```

Note earlier diversification initially appears slow, and then accelerates rapidly between 60 MYA–70 MYA (fig. 8.10). If we think back to the phylogeny of Australian mammals, the first evolutionary split approximately 165 MYA was between monotremes and all other mammals. The second split represents the separation between marsupials and placental mammals around 147 MYA, although recent evidence suggests the actual split may have preceded this date by several million years (Graves and Renfree 2013). The appearance of initially slow diversification can in part be explained by the sampling in our tree, which omits many of the early mammal diversification events that occurred on other continents.

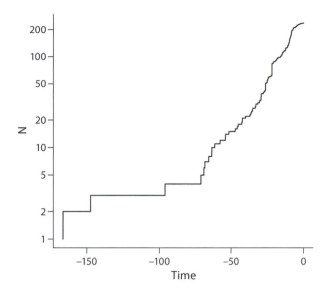

Figure 8.10. Lineage, or diversification, through time (LTT) plot for Australian mammals.

For example, of the native Australian fauna, there are no representatives of Carnivora, Artiodactyla, or primates; these are all clades that diversified extensively elsewhere. From 60 MYA to approximately10 MYA the LTT plot appears to be almost linear; because the y-axis is on a logarithmic scale, this actually translates to an exponential diversification process as might fit a pure-birth model of evolution (see below). Last, from 10 MYA to the present we can observe an apparent slowdown in diversification.

Now let us compare expected LTT plots under different evolutionary scenarios. First, we can examine a pure-birth model of clade growth, also called a Yule model, and compare it to a model with high extinction (fig. 8.11) using functions from the *phytools* library:

```
b.tree<-pbtree(b=0.1, d=0, n=250)
ltt.plot(b.tree, log = "y")
title("Lineages Through Time Plot with B = 0.1 D = 0")

bd.tree<-pbtree(b=0.1, d=0.08, n=250, extant.only=T)
ltt.plot(bd.tree, log = "y")
title("Lineages Through Time with B = 0.1 D = 0.08")
```

To the left of figure 8.11 we can see the expected LTT plot assuming zero extinctions (D = 0). There is some variability when diversity is low initially, which is to be expected when modeling a stochastic process. However, when the number of lineages >20, the LTT plot is almost linear, matching our phylogeny of Australian mammals. In the plot to the right of figure 8.11, with nonzero extinction, we see that early rates appear quite rapid. This is fairly common among empirical data sets. One possibility is that for many named clades early diversification rates are in fact unusually rapid, as might be predicted in, for example, an adaptive radiation, and then rates slow down through time (see Phillimore and Price 2008, Rabosky and Lovette 2008). However, because we specified the model of diversification ourselves, we know that rates have not changed through time, but the probability

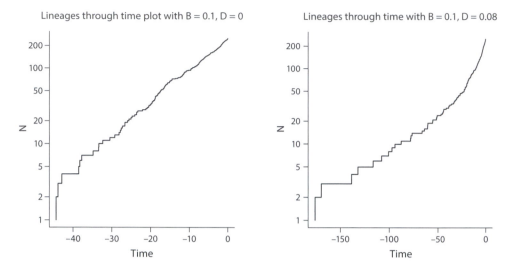

Figure 8.11. Lineage through time (LTT) plots for simulated data assuming a pure-birth model (*left*) and a birth-death model (*right*) of clade growth.

of speciation and extinction has been constant. In this case, the appearance of rapid early rates may be explained by a selection effect, whereby trees that have slow diversification initially (by chance) tend not to survive to the present because they spend a longer time at low diversity, and thus when extinction rates are relatively high there is an increased chance that all lineages wander to extinction. This selection effect can be demonstrated by running the `pbtree()` function multiple times with high relative extinction. Many simulations will result in the following error:

```
Warning:
    no extant tips, tree returned as NULL
```

This indicates that all species went extinct before the simulation ran to completion. Thus the observed trees will more often tend to have, by chance, diversified rapidly early on. In the birth-death LTT plot we also see an uptick toward the present, and this again represents a sampling effect, rather than a change in rates. Because lineages have to first come into existence before they can become extinct, toward the present we therefore increasingly sample recently diverged lineages upon which extinction has not yet had time to act. This phenomenon has been called the "pull of the present" (Nee et al. 1994). Note that this is distinct from a similar term—"pull of the recent"—that derives from preservation bias in the fossil record, where we have better records for more recently preserved taxa (Raup 1979).

For each tree, we can use the distribution of branching times to back calculate the speciation and extinction rates using the `bd()` function in the *laser* library. First, let's estimate rates for the simulated trees because we know the true parameters under which they were generated (keep in mind that because the tree simulations were stochastic, model parameters will differ each time we run the code):

```
b.times<-branching.times(b.tree)
bd(b.times, ai = c(0.1, 0.5, 0.9))
```

```
$LH
[1] 330.8867
$r1
[1] 0.094853
$a
[1] 0.229227
$aic
[1] -657.7735

> b.times<-branching.times(bd.tree)
> bd(b.times, ai = c(0.1, 0.5, 0.9))
$LH
[1] 173.3604
$r1
[1] 0.02301096
$a
[1] 0.7801363
$aic
[1] -342.7209
```

For each tree, we have the log likelihood (LH) and AIC (aic) of the fit of the model (see below), along with the maximum likelihood estimate of the speciation rate (r1) and relative extinction rate (a). For the pure-birth tree (b.tree) we know that r1 = 0.1 and a = 0. Our model fit estimates r1 = 0.09 and a = 0.2, which seems reasonably close, but the extinction fraction is overestimated. Perhaps this is because we are over-fitting our model, and we do not need the extinction parameter at all. We can compare the fit of the birth-death model to the pure-birth model by comparing AIC scores using fitdAICrc():

```
fitdAICrc(b.times, modelset = c("pureBirth", "bd"))

--------------Model Summary----------------
MODEL pureBirth
Parameters: r1
LH 330.1844
AIC -658.3688
r1 0.1087058
a -658.3688


--------------------------

MODEL bd
Parameters: r1, a
LH 330.8867
AIC -657.7735
r1 0.094853
a 0.229227


--------------------------
Best Constant Rate Model = pureBirth AIC -658.3688
```

We can now see the pure-birth model is marginally favored by AIC (it has a more negative value), and under this model r1 is estimated at 0.11, fitting closely to the true underlying model.[2]

For the birth-death tree, we know that r1 = 0.1, and a is the relative extinction fraction calculated as B/D = 0.08/0.1 = 0.8. Our model fit estimates r1 = 0.02 and a = 0.8, indicating that we do well at estimating extinction, but less well at correctly estimating speciation on this particular tree (again results will differ for each simulated tree). Nonetheless, when we compare the fit of the pure-birth model to a birth-death model, the latter is strongly and correctly favored by AIC (ΔAIC = 42.83):

```
> fitdAICrc(bd.times, modelset = c("pureBirth", "bd"))

--------------Model Summary----------------
MODEL pureBirth
Parameters: r1
LH 150.9427
AIC -299.8853
r1 0.05276753
a -299.8853


--------------------------

MODEL bd
Parameters: r1, a
LH 173.3604
AIC -342.7209
r1 0.02301096
a 0.7801363


--------------------------

Best Constant Rate Model = bd AIC -342.7209
```

Now let us compare the fit of diversification models in our Australian mammal tree:

```
times<-branching.times(phy)
fitdAICrc(times, modelset = c("pureBirth", "bd"))

--------------Model Summary----------------

MODEL pureBirth
Parameters: r1
LH 10.29293
AIC -18.58585
r1 0.04669079
a -18.58585


--------------------------
```

2. Note that in this release, the output of the function has a small bug and reports the AIC as the relative extinction rate in the pure-birth model.

```
MODEL bd
Parameters: r1, a
LH 10.5542
AIC -17.10841
r1 0.04171677
a 0.1823903

--------------------------

Best Constant Rate Model = pureBirth AIC -18.58585
```

A pure-birth model is favored. From our LTT plot and knowledge of the history of Australian mammals, this result might seem surprising, and certainly our LTT plot does not resemble that from our pure-birth simulations. Instead we see initially slow rates, followed by a rapid increase in diversification rates, and then a slowdown toward the present. The recent slowdown is opposite to expectations from the birth-death model, but rates do not appear very constant through time. We can therefore evaluate a wider range of diversification models using functions from the *laser* library, including a multirate Yule model (`yule-n-rate()`), which allows shifts in rates at some time *st*, and a diversity dependent model (`DensityDependent()`), in which speciation rate is a function of the number of extant lineages at any point in time relative to some carrying capacity, K:

```
density.dependent<-DDL(times)
density.dependent
$LH
[1] 10.29287
$aic
[1] -16.58574
$r1
[1] 0.04669179
$kparam
[1] 2299063
```

We find no better fit of a density-dependent model to the pure-birth model (AIC = −16.59 vs. −18.59). Now let's evaluate the fit of a multirate model:

```
two.rate<-yule2rate(times)
two.rate
        LH          r1          r2          st1          AIC
19.13060287  0.04744460  0.02070393  0.60000000  -32.26120574
```

This model is greatly favored over the single-rate Yule model (AIC = −32.26 vs. −18.59), with the two rate classes (`r1` and `r2`) varying by over a factor of two. Moreover, adding in additional rate classes provides an even better fit:

```
three.rate<-yule3rate(times)
three.rate
            LH              r1              r2              r3
7.401340e+01  5.069878e-02  2.277306e+11  1.399580e-02
            st1             st2             AIC
2.700000e+00  2.700000e+00  -1.380268e+02
```

We can additionally explore sequentially 3-, 4-, and 5-rate class models, and in general it appears that the more rate classes the better fit of the model; although note that computational times can be excessive when fitting multiple-rate models. Given that our phylogeny represents a phylogenetically nonrandom sample of mammal diversity, it is not surprising that we reject the single-rate model. As discussed above, our sampling is highly uneven across taxonomic depths, and this could bias diversification rate analysis. Nonetheless, there is some evidence from the complete mammal phylogeny to suggest that early net diversification rates (speciation rate minus extinction rate) were in fact relatively slow (Bininda-Emonds et al. 2007). Perhaps more surprising is that we did not find support for a density-dependent model, despite the fact that we can observe in our LTT plot an apparent slowdown toward the present. The *laser* library also allows us to explore the distribution of branch lengths through time using Pybus and Harvey's gamma statistic, which describes the average weighting times to speciation, relative to the midpoint of the tree (Pybus and Harvey 2000). Negative values indicate that branching has been more rapid earlier in the tree (an evolutionary rate slowdown), whereas positive values indicate that branching rates increase toward the tips of the tree (evolutionary rate increase):

```
gammaStat(phy)
[1] -9.847542
Warning messages:
1: In (2:N) * g :
longer object length is not a multiple of shorter object length
```

Note that the function returns several warning messages; this is because our tree is not fully resolved. We can arbitrarily resolve polytomies using zero branch lengths with the `multi2di()` function in *ape*, but consider whether this is a sensible step in your own tree. Arbitrarily resolving nodes with zero branch lengths will tend to overestimate tip branches, introducing a negative bias into the gamma statistic. Alternative methods allow us to generate a distribution of fully resolved trees with more meaningful branch lengths, but these tend to impose an equal-rate branching process within unresolved clades, introducing a separate bias (Kuhn et al. 2011). For convenience, we can stick with our quick fix:

```
gammaStat(multi2di(phy))
[1] -2.657112
```

The gamma statistic is negative, supporting our observation that branching rates show a decrease toward the present (bias introduced by polytomies notwithstanding). We can formally test whether this trend is significant using the matrix of branching times from the resolved tree in *laser*:

```
gamStat(branching.times(multi2di(phy)))

------------------------------

Calculated gamma: -2.657112
pvalue: 0.003940658
test: one-tailed; Ho: rates have not decreased over time
*assumes complete taxon sampling.

$gamstat
[1] -2.657112
```

```
$pval
[1] 0.003940658

$test
[1] "one-tailed; Ho: rates have not decreased over time"
```

These results support our observation that there is a significant trend for a slowdown in diversification toward the present ($P = 0.004$). What might explain the shape of our LTT plot? Once again, taxonomic sampling might be part of the explanation. While our representation of species diversity within clades at shallow depth in the phylogeny might be relatively complete, there can be a lag in the formation of new species and our ability to formally diagnose them as such. For example, recent speciation events might split lineages that remain morphologically similar, or the time for speciation itself might be protracted (Etienne and Rosindell 2012). We thus underestimate current species richness, leading to an undercount of extant lineage diversity, and the apparent downturn in LTT plots (Etienne and Rosindell 2012).

The *laser* library provides functions that allow us to also fit models of increasing or decreasing diversification through time (fitSPVAR(), fitEXVAR(), and fitBOTHVAR()). Temporal trends toward lower diversification might be explained by niche filling or other diversity dependent processes (Phillimore and Price 2008, Rabosky and Lovette 2008). Shifts to increased diversification might be attributed to ecological release, as has been proposed to explain the increased diversification of mammals in the Paleocene (Alroy 1999), but see Bininda-Emonds et al. (2007); alternatively, fluctuations in environment could explain both diversification increases and decreases. Another explanation for rate variation is the evolutionary origin of key innovations; that is, biological traits that open up new adaptive zones (Van Valen 1971). In mammals, various traits such as hair, mammary glands, and dentition have been suggested as putative key traits that influenced diversification (e.g., Woodburne et al. 2003). Because key innovations characterize clades rather than time periods, such traits might better explain rate variation among lineages, rather than trends through time. For example, most phylogenetic trees demonstrate high imbalance—the disparity of taxonomic richness between sister clades. Because sister clades are, by definition, the same age, differences in richness must be attributed to differences in net diversification rates (Barraclough, Nee, et al. 1998). High imbalance might be expected if a key innovation occurs in the stem lineage for one clade, but not its sister.

Colless' I statistic (*Ic*) is one of various tree shape metrics that calculates a simple measure of tree imbalance. The *Ic* index computes the sum of differences in richness between sister clades, and can be calculated using the *apTreeshape* library:[3]

```
i.tree<-as.treeshape(phy, model="yule")
colless(i.tree)
[1] 1444
```

The raw index is not especially meaningful because we would expect it to be larger in trees sampling more taxa, irrespective of the tree topology unless trees are completely

---

3. Note here we first convert our *phylo* object into a *treeshape* object specifying a Yule model to resolve polytomies.

symmetrical. We can therefore standardize the index to expectations derived from an equal-rate Yule process:

```
colless(i.tree, norm = "yule")
[1] 1.710888
```

Positive values indicate that the tree is relatively imbalanced in comparison to expectations derived from an equal-rate Yule process. Matching a general trend across most empirical phylogenies, the tree for Australian mammals appears highly imbalanced. We can test for significance using Monte Carlo simulations to generate a distribution of expected Ic values assuming a Yule branching process:

```
colless.test(i.tree, model = "yule", alternative =
        "greater", n.mc = 999)
        Test of the yule hypothesis using the Colless index
Statistic = 1444
Standardized Statistic = 1.710888
p-value = 0.02002002
alternative hypothesis: the tree is less balanced than pre-
        dicted by the yule model
```

Results confirm that the tree for Australian mammals is significantly more imbalanced than expected from an equal-rate model. Diversification rate shifts might thus be frequent across the tree, but not necessarily demonstrate a temporal trend. The medusa() algorithm in *geiger* allows us to explore the location of rate variation across clades, rather than over time. The algorithm first fits a single diversification model to the entire tree, and then sequentially adds breakpoints in the diversification process such that rates vary across the tree. Here we again focus on the simple Yule model, although birth-death and mixed models can also be explored:

```
sp.rich<-data.frame(phy$tip.label, rep(1, length(phy$tip.
        label)))
names(sp.rich)<-c("taxon", "n.taxa")

med1<-medusa(phy, richness = sp.rich, model = "yule", crite-
        rion = "aic")
print(med1$summary)
```

	partitions	split	cut	lnL	k	aic	aicc
1	1	NA	node	−699.6991	1	1401.398	1401.408
2	2	291	node	−675.2406	3	1356.481	1356.542
3	3	254	node	−665.1195	5	1340.239	1340.392

Here we can see that two rate shifts are favored. The addition of a third rate shift does not further decrease AIC. The medusa algorithm is attractive because it additionally identifies for us the nodes at which inferred shifts are likely to have occurred, and it is able to handle incomplete phylogenies, assuming the true diversity of extant tips is known (richness =); it does this by providing a two-column table with column names "taxon" and "n.taxa" representing taxon names and species richness, respectively. Here we have generated a table (sp.rich) setting richness equal to 1 for all tips. We can visualize the

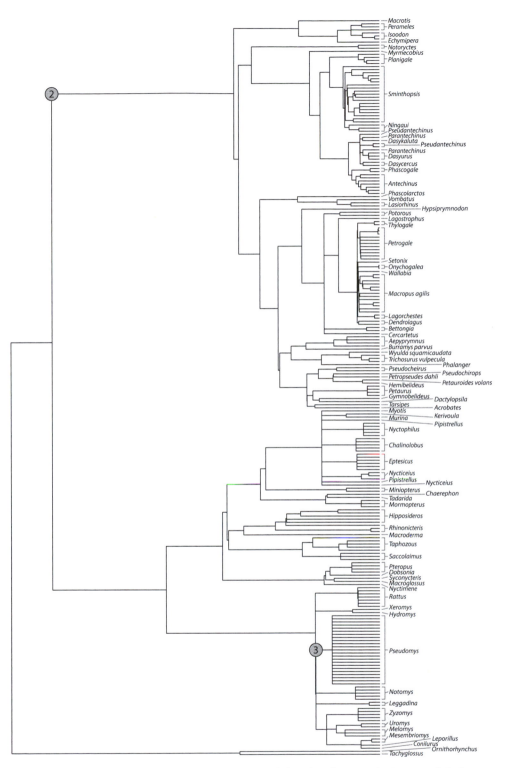

Figure 8. 12. Location of diversification rate shifts (*2* and *3*) within the Australian mammal phylogeny (see text).

location of rate shifts using `plot` on the summary of the medusa output (fig. 8.12), specify-ing the AIC criterion to select the appropriate number of regime shifts, and some standard plot options (`cex` and `label.offset`) to make the figure display nicely:

```
plot(summary(med1, criterion = "aic"), cex=0.35,label.
    offset=1, edge.width=1.5)
```

The first rate shift (2) is to a higher rate, and represents the diversification of marsupials. The second rate shift is to a lower rate (3) and highlights the genus *Pseudomys*, a clade of Australian mice and one of the few placental lineages that likely diversified within Australia. However, on closer inspection we can see that this clade represents a single large polytomy. Probably at the time of tree reconstruction there were was no molecular phylogeny for the genus, and so we have information based only on taxonomy, and not on evolutionary rela-tionships between species. Because we push back all divergences within unresolved clades to their most recent common ancestor, we lose information on branching times within the clade, and thus we should not have much faith in encompassed rate estimates. If we look carefully at the LTT plot above, we can see the influence of this polytomy as a rapid jump in diversity approximately 20 MYA. Finally, we need to remind ourselves that our phylogenetic tree is highly incomplete for placental mammals; thus, the apparent upshift in evolutionary rates within marsupials might simply represent the more complete sampling of diversity in this clade.

The analysis of evolutionary rate shifts is constantly improving, and more sophisticated methods are becoming available. One recent example is BAMM (Bayesian analysis of mac-roevolutionary mixtures, http://bamm-project.org/introduction.html), a program written in C++ that uses reversible jump Markov chain Monte Carlo to evaluate the large parameter space of possible models (Rabosky et al. 2013, Rabosky 2014). Although analyses are run outside of R, and we therefore do not provide details here, post-run analysis and visualiza-tion is performed using the R package BAMMtools. Additional tools allow us to explore shifts in evolutionary rates associated with the gain and loss of particular character states (e.g., `BiSSE()` and `multiSSE()`) or geographical shifts (`GeoSSE()`) in the *diver-sitree* library. These methods have the advantage of allowing for simultaneous estimates of both speciation and extinction rates, as well as rates of evolutionary (or geographic) transi-tions. This simultaneous estimation is particularly important for correctly inferring rates of state change because a character state associated with high extinction has less probability of being observed among extant taxa, even if shifts to that state are frequent (Maddison 2006, Maddison et al. 2007). Ignoring extinction therefore would bias us to infer a lower rate of transitions to that particular character state and of course the opposite is true for states associated with lower extinction or higher speciation. A more detailed description of these approaches and their implementation in R is provided by R. Fitzjohn (2010, 2012), but see also a recent critique of this method by Rabosky and Goldberg (2015).

8.3. CONCLUSION

We have explored here macroevolutionary trends across space and through time. We have shown how phylogenetic diversity covaries closely with taxonomic richness, but that it is possible to identify regions that contain more or less evolutionary history than expected from the number of taxa found within them. Such regions might represent centers of recent rapid diversification (containing less evolutionary history than predicted) or museums of diversity (containing more evolutionary history than predicted). Variation in diversification rates can be explored directly using the distribution of branching times on the phylogeny, and we have reviewed several approaches for mapping diversification rates (or some proxy for rates) across space, as well as methods for identifying shifts in rates on the phylogenetic tree itself. Additional methods are available to test hypotheses relating rate shifts to particular biological traits or geographic distributions, although we do not provide details here. Importantly, in all analyses of diversity taxonomic sampling and tree resolution have to be carefully considered; although various methods are available to account for missing species, the true diversity of tip taxa must be known with confidence, and rate estimates are less reliable than if we have a completely sampled tree. In our example, we used the phylogeny of Australian mammals. While the isolation of the continent represents unique opportunities to study evolutionary process, we included a number of taxa that likely evolved elsewhere and colonized via long distance dispersal, and our analyses of diversification thus misses much of Australian mammal evolution.

◇◇◇

CHAPTER 9

Using Phylogenetic Information to Make
Better Conservation Decisions

Study the past if you would divine the future.
—*Confucius*

There is a growing weight of evidence that indicates we may be entering a mass extinction event. Species are going extinct at rates much faster than background rates estimated from the fossil record, and perhaps faster by over an order of magnitude (Pimm et al. 1995). Projections for the future are even more alarming as rates are predicted to increase even further unless we take action to reduce them (Mace et al. 2005). We have transformed the terrestrial environment, which is now dominated by people (Vitousek et al. 1997), and human consumption may account for one-quarter of global productivity, the amount of the annual total plant growth (Haberl et al. 2007). The main recent and current drivers of extinction are well known and include habitat loss and fragmentation; however, additional causes include pollution, overharvesting, invasive species, and disease. We continue to transform the environment, and we have now also influenced climate. Some estimates suggest that human-caused climate change is likely to be the dominant cause of extinction in the near future, with up to 37% of species projected to be committed to global extinction by 2050 (Thomas et al. 2004), although this number depends on the future climate scenarios considered and varies with assumptions regarding species movements in response to such change. The International Union for Conservation of Nature (http://www.iucnredlist.org/) has documented the slide toward extinction of many groups in its Red List publications. In mammals, 21% of species are currently listed as threatened with extinction and similar proportions seem threatened in other taxonomic groups. Global biodiversity may thus change considerably in our lifetime unless action is taken to reduce the rate at which species are being lost.

Conservation funding remains far below that required even to manage the current areas set aside for protection (James et al. 2001, Halpern et al. 2006). It is an unfortunate reality, therefore, that we have to prioritize among areas for conservation if we are to maximize the return on our efforts (Myers et al. 2000). The largest and best-funded conservation effort to date is the Biodiversity Hotspots initiative (Myers et al. 2000, Mittermeier et al. 2004, Brooks 2006, Zachos and Habel 2011), which has been supported by Conservation International (http://www.biodiversityhotspots.org). This effort identifies areas that are particularly rich in species, especially endemic species, and are currently under threat from habitat loss (Myers et al. 2000). However, the criteria required for qualifying as a hotspot for conservation has been much debated, and requires that different dimensions of diversity are weighted and contrasted against each other. Most usually, hotspot approaches are species-centric, and there are many reasons why this might seem sensible: species are (mostly) easily quantifiable units, we can easily compare species richness of taxa and areas, and species have obvious appeal to the nonscientist. Implicit within this framework, however, is an assumption that all species are equal—but as this book attests, evolution suggests otherwise. Biodiversity was defined at the 1992 United Nations Conference on Environment and Development in Rio de Janeiro as "the variability among living organisms from all sources, including, 'inter alia,' terrestrial, marine and other aquatic ecosystems, and the ecological complexes of which they are part: this includes diversity within species, between species and of ecosystems." If our aim is to maximize the preservation of biodiversity *sensu lato*, then conservation must think beyond species. Interestingly, in his seminal book *The Diversity of Life*, E. O. Wilson (1992) expanded this definition of the variability among organisms to explicitly include higher taxa. Higher taxa, however, are somewhat subjective units, and it is not clear what defines a genus or a family rank; phylogeny provides a more objective tool for incorporating evolutionary history into conservation prioritization.

The evolutionary history of most species (notable exceptions being asexual and parthenogenic organisms) can be described as a series of nested relationships, most typically represented as their phylogeny, but also often captured by taxonomy. We can therefore group closely related species into evolutionary clades or taxa that share much of their evolutionary history, and hence life histories and ecologies (see chapter 1). Species falling in the same clade will thus be more similar, on average, than species in different clades, as we have discussed in the preceding chapters. For example, consider two species of grass (Poaceae) and two species of daisy (Asteraceae). The two grass species are more likely to share similar leaf morphologies (typically alternate, distichous [in one plane], and with parallel veins), life histories (annual or perennial), and pollination mode (wind) among other traits. So too will the two daisies (e.g., lobed leaves; annual, biennial, or perennial habit; insect pollination; hardy seed bank, etc.). As a result of such shared traits, closely related species are likely to have similarly ecologies. The two grass species are likely to support similar herbivores adapted to high silica herbage, and to build up combustible leaf litter (D'Antonio and Vitousek 1992). The result is a very different ecosystem where the two grasses dominate compared to where the daisies are dominant. This link between phylogenetic distance and ecological similarity underpins much of ecophylogenetic analyses, but it is also important when we consider the conservation of ecosystems.

If our goal is to maximize biodiversity (rather than just species richness), one approach would be to measure all the various traits for each species, and select the set of species that maximizes trait diversity. However, such an effort would require vast resources and many hours of labor; fortunately, we can use a simple proxy—phylogeny. Phylogenetic metrics, much like those discussed in chapter 3, include a suite of flexible, low-dimension measures that are relatively inexpensive to estimate (Faith 1994, Crozier 1997, Faith 2002). Since we can assume that phylogenetic relatedness should be correlated with species morphology and ecology, maximizing the preservation of phylogenetic diversity will also tend to maximize the preservation of phenotypic and ecological diversity. Phylogeny can therefore provide a powerful tool for aiding conservation decision making.

9.1. WHY PRESERVE EVOLUTIONARY HISTORY?

The Tree of Life (ToL), the phylogenetic tree connecting all living organisms, provides a powerful metaphor for conservation biology. Speciation, which occurs over thousands to tens of thousands of years, slowly adds new shoots; while extinction, occurring on the order of tens to hundreds of years, is systematically pruning the branches from the tips of the tree, and with them the unique evolutionary histories they represent. Further, risk of extinction is frequently associated with particular biological traits, such as large body size and low reproductive rate (Purvis, Apagow, et al. 2000, Purvis, Gittleman, et al. 2000, Cardillo et al. 2005); therefore, not only are we thinning the branches, but sometimes we are losing whole limbs from the tree and this seems exacerbated by a tendency for the most threatened species to fall within species-poor clades (Russell et al. 1998, Purvis, Apagow, et al. 2000).

Reasons for preserving evolutionary history are many; in particular, Mooers et al. (2005) suggest that a time-calibrated estimate of phylogenetic diversity (1) provides a useful proxy for accumulated trait change across multiple traits, (2) allows a common scale to be applied across taxa, and (3) is an easily comprehensible metric for policy making and public understanding. As policy makers and educators, we might be more interested in the last argument. Evolutionary history might be considered a valued part of our natural heritage, for which we act as both guardians and docents, and equivalent to the value represented by diverse human cultures. As ecologists and conservation biologists, our focus might be on the first two arguments. First, if phylogenetic diversity captures ecological and functional divergence, as we lose the branches on the tree of life, we also lose the biological diversity necessary for ecosystem function; that function provides us with valued ecosystem services, such as clean water, flood control, crop pollination, and so forth. To date, the relationship between phylogenetic diversity and ecosystem function has been little studied; however, recent evidence suggests that evolutionary history might be more important than species richness in explaining productivity-diversity relationships (Maherali and Klironomos 2007, Cadotte et al. 2008, Cadotte 2013), while also encompassing the diversity of important functional traits (Forest et al. 2007, Bässler et al. 2014).

As rates of extinction increase, the ToL has come under sustained and increasing attack. However, simulations suggest the impact of extinction on the loss of phylogenetic diversity (PD; see chapter 3) may be inflated, and that much of the tree of life might remain even after a devastating extinction event with over 80% of PD remaining, even when 95% of species are lost (Nee and May 1997). However, this model assumed that extinction strikes taxa at random, referred to as the "field of bullets" scenario (Raup et al. 1973), in which survival is not contingent upon inherent species' attributes, but is simply down to chance—that is, the bullets are not targeted. Further, Nee and May evaluated the expected loss of PD from coalescent trees, which tend to be more topologically balanced than observed phylogenetic topologies (Heard 1992). There is accumulating evidence indicating both these assumptions (random extinctions and balanced trees) are violated. Empirical data suggests phylogenetic nonrandomness in extinction risk is a common feature in many taxa, including mammals (Purvis, Agapow, et al. 2000), flowering plants (Vamosi and Wilson 2008), birds (Russell et al. 1998, Von Euler 2001, and amphibians (Cooper et al. 2008). The loss of PD under phylogenetically nonrandom extinction may be greater because we lose not only the unique evolutionary branches from which extinct species descend, but also the network of branches that form connections among them (Purvis, Agapow, et al. 2000). Furthermore, greater topological imbalance also elevates PD loss, even under random extinction, because of greater frequency of species-poor clades with long, distinct evolutionary histories (Heard and Mooers 2002). When the topology of real phylogenetic trees are examined, they tend to be much more imbalanced than expected (Heard 1996, Aldous 2001). The actual loss of evolutionary history under projected extinction scenarios may, therefore, be much greater than suggested from the rather optimistic scenario presented in Nee and May's simulations.

Several studies have attempted to quantify the expected loss of PD if currently threatened species become extinct. Purvis, Agapow, et al. (2000) used IUCN Red Listings for
mammals and birds, and showed greater than predicted loss of PD with increasing extinction intensity (removing from the phylogeny successively lower IUCN-rated species). Similar trends were shown in other clades, including plants (Vamosi and Wilson 2008, though
the opposite has also been observed [Davies et al. 2011]) and suggest a large proportion of
our evolutionary heritage is currently under threat. Why might extinction risk cluster on
phylogenies and why are the most evolutionary distinct species frequently those at greatest
risk of extinction? At first, it might seem counterintuitive that a trait that confers high risk
of extinction should be heritable. Would not such a trait quickly be removed by natural
selection? However, comparative studies have shown that risk is strongly correlated with
biological and ecological traits that covary with phylogeny, such as geographic range, body
size, and reproductive rate (Pacala and Rees 1998, Owens and Bennett 2000, Purvis, Gittleman, et al. 2000, Cardillo et al. 2005, Sodhi et al. 2008, Cooper and Purvis 2010). Such traits
might confer selective advantage to individuals by, for example, providing greater parental
investment to offspring or increased competitive ability for limited resources, while at the
same time consigning lineages to extinction.

Identifying correlates of extinction risk is a useful tool for conservation biology, allowing the identification of vulnerable species—those that have traits that predispose them to
high risk but are not yet threatened—offering a chance for pre-emptive conservation (Cardillo et al. 2006). However, more recent simulations have demonstrated that phylogenetic
structure in extinction risks explains only very little of the large loss of evolutionary diversity predicted by the current extinction crisis (Heard and Mooers 2000, Huang et al. 2011,
Parhar and Mooers 2011). Contrary to previous speculation (Purvis, Agapow, et al. 2000,
Corey and Waite 2008), it would appear that the shape of the tree may therefore be more
important in determining the magnitude of PD loss than the distribution of extinction
risk on the tips. Again using simulations, Heard and Mooers (2000) were able to show how
phylogenetic topology influenced PD loss. As predicted, more imbalanced trees resulted in
greater loss of PD, but the relationship is also weak; rather, the best topological predictor of
PD loss is tree "stemminess"—the relative distribution of branching events from the root
to the tip of the tree. We can explore the relationship between tree stemminess and PD loss
using simulations, first generating a random coalescent tree (`rcoal()`), as we have done
before, and pruning tips to examine the loss of PD. We then simply need to transform this
tree topology to stretch either internal or terminal branches using Pagel's δ (see chapter 4).
In the following code we generate a simple "for" loop to run over 100 trees, and we use two
values of δ: $\delta = 0.1$ stretches internal branches (the tree is considered stemmy), and $\delta = 10$
stretches terminal branches (the tree is considered tippy). We then simulate extinction by
removing 30 species at random from the tree, and quantify the proportional loss of PD for
each topology (fig. 9.1):

```
PD.loss<-NULL
PD.loss.delta0.1<-NULL
PD.loss.delta10<-NULL

for (i in 1:100){
tree<-rcoal(64)
extinct<-sample(tree$tip.label, 20)
pruned.tree<-drop.tip(tree, extinct)
```

```
PD.loss[i]<-(sum(tree$edge.length)-sum(pruned.tree$edge.
        length))/sum(tree$edge.length)

delta0.1.tre<-rescale(tree, model = c("delta"), 0.1)
pruned.delta0.1.tre<-drop.tip(delta0.1.tre, extinct)
PD.loss.delta0.1[i]<-(sum(delta0.1.tre$edge.length)-
        sum(pruned.delta0.1.tre$edge.length))/sum
        (delta0.1.tre$edge.length)

delta10.tre<-rescale(tree, model = c("delta"), 10)
pruned.delta10.tre<-drop.tip(delta10.tre, extinct)
PD.loss.delta10[i]<-(sum(delta10.tre$edge.length)-
        sum(pruned.delta10.tre$edge.length))/sum(delta10.
        tre$edge.length)
}

results<-cbind(PD.loss.delta0.1, PD.loss, PD.loss.delta10)
boxplot(results)
```

Our simulations demonstrate clearly that the proportional loss of PD is much greater for the more tippy trees (fig. 9.1). However, most likely, it is the interaction between tree shape and phylogenetic structure in extinction risk that determines the magnitude of PD loss.

The true functional cost to the loss of PD in natural systems remains to be evaluated; however, there are strong theoretical justifications for preserving PD, as we have described above. It would seem a sensible approach to apply the precautionary principle. We are therefore faced with new challenges of how best to select the set of species or areas to maximize the preservation of evolutionary heritage. In the previous chapter, with our exploration of geographical patterns in mammal evolutionary history across Australia, we presented one simple illustration of how sites could be selected to maximize total phylogenetic diversity;

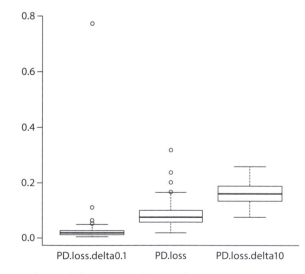

Figure 9.1. Proportional loss of phylogenetic diversity from a 64-species tree in which 30 species go extinct for different δ transformations (0.1 and 10 representing "stemmy" and "tippy" trees, respectively). The middle plot represents the untransformed tree (δ = 1).

we now expand on this theory in the following sections, considering the alternative dimension of phylogeny.

9.2. QUANTIFYING EVOLUTIONARY HISTORY

The early 1990s saw the rise of phylogenetically based metrics for conservation prioritization (Vane-Wright et al. 1991, Faith 1992a). From this inception, phylogenetic metrics were either used to compare places or species. Below, we discuss these approaches and advances that incorporate geographical rarity and range sizes.

9.2.1. Using PD to Evaluate Sites

In an earlier chapter (chapter 3), we calculated PD as the sum of the edges for a community phylogeny, but the first incarnations of PD were to evaluate the relative conservation value of sites (Faith 1992a, 1992b, 2002; Forest et al. 2007; Cadotte and Davies 2010). Faith's PD (1992a) was explicitly designed to compare the proportional representation of a regional phylogeny within sites (fig. 9.2). In a conservation context, we thus suggest the root node should be retained in the calculation of the site PD even if the species sampled at a particular site do not span the tree root.

There are a number of ways one could calculate both types of PD in R. As we have already seen (chapter 3), there is the `pd()` function in *picante* that will calculate both of these across a number of sites. As a reminder, this function takes three arguments: a site by species community matrix, a phylogenetic tree representing these species, and an argument to include the root. To run it, including the root, simply set the argument `include.root` to TRUE:

```
library(picante)
data(phylocom)
pd(phylocom$sample,phylocom$phylo,include.root=TRUE)
```

This results in:

	PD	SR
clump1	16	8
clump2a	17	8
clump2b	18	8
clump4	22	8
even	30	8
random	27	8

To exclude the connection to the root when sites do not all contain species that traverse the root in the phylogeny, simply set `include.root` to FALSE. This results in:

	PD	SR
clump1	14	8
clump2a	16	8
clump2b	18	8
clump4	22	8
even	30	8
random	27	8

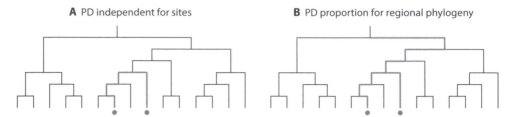

A PD independent for sites   **B** PD proportion for regional phylogeny

Figure 9.2. The branches of a regional phylogeny represented by species within a single site. The first example (*a*) sums only the branches connecting resident species directly, and the second (*b*) includes the connection to the root.

Notice that PD is lower in two of the sites (`clump1` and `clump2a`), meaning that these sites did not contain species that spanned the root.

In effect, when `include.root` is `FALSE` a subtree is created for each community and branch lengths are summed from the vector of edge lengths. We will use a function we name `subTree` to demonstrate:

```
subTree<-function(phy,species) {
        all.in<-is.na(match(species,phy$tip.label))
        #first we check to make sure all species are in the
            phylogeny
        if (sum(all.in)>0)
            paste("species",as.character(species[all.in]),
             "not in phylogeny")
        #this tells you which species are not in phylogeny
        dropme<-phy$tip.label[!phy$tip.label %in%
            unique(species)]
        #this identifies the species not found at a site
        sub.tr<-drop.tip(phy,dropme)
        #this uses the drop.tip function in ape to remove
            species
        }
```

We can run this on "clump1" from the *phylocom* data in *picante* by creating a subtree and summing the edge lengths:

```
site.tree<-subTree(phylocom$phylo,names(phylocom$sample[1,-
    phylocom$sample[1,]>0]))
```

and PD is simply calculated as:

```
sum(site.tree$edge.length)
```

This gives 14, same as the `pd()` function. To retain the root node when calculating PD in this way is not straightforward. *Picante* simply adds the difference between the maximum depth of the full and subtrees. Luckily, *picante* includes a useful function call `node.age()`, which tells you how far back each node is in the phylogeny by summing the descending edges. Here is a simplified version of the calculation:

```
#to keep root node distance add difference between max depth
    of full and subtrees
site.tree.depth<-max(node.age(site.tree)$ages)
full.tree.depth<-max(node.age(phylocom$phylo)$ages)
#and so PD is simply
sum(site.tree$edge.length) + (full.tree.depth-site.tree.
    depth)
```

This gives 16, confirming the `pd()` function result with `include.root=TRUE`.

9.2.2. Calculating ED to Compare Species

The evolutionary distinctiveness of a species describes its relative "uniqueness" within a phylogeny, and can be measured as the number of evolutionary divergences (evolutionary splits or nodes) from the root to the respective tip in the phylogenetic tree (Vane-Wright et al. 1991); see figure 9.3A. Species characterized by many splits from root to tip have many close relatives and are less evolutionary distinct. Vane-Wright et al. (1991) provided the first quantitative valuation of taxonomic distinctiveness (TD) for conservation—the formulation here follows Redding et al. (2008):

$$TD(T,i) = \frac{1}{\sum\limits_{v \in q(T,i,r)} \deg_{out}(v)} \tag{9.1}$$

where, for a tree T, the set $q(T,i,r)$ includes the node splits between species i and the root of the tree, r. The value of $\deg_{out}(v)$ for any node is 2 in a perfectly resolved bifurcating tree and >2 for a node containing a polytomy subtended by more than two lineages.

From equation 9.1, a species with few preceding splits is taxonomically (phylogenetically) more distinct than one within a more diverse clade. However, TD it is sensitive to the taxon sampling of the phylogeny and does not include information on branch lengths. The addition of branch lengths makes distinctiveness measures less sensitive to taxon sampling because estimates of time since divergence are not strongly influenced by the exclusion of sister taxa, whereas the number of evolutionary splits is obviously highly dependent upon the number of included species. Two derivations of TD consider branch lengths in the calculation of distinctiveness. The first, Equal Splits (ES; Redding and Mooers 2006, Redding et al. 2008) divides a branch length by the number of lineages originating from the node directly below it (fig. 9.3B). This formulation, similar to TD, scales the length of the branch, λ_e, by the number of splits preceding node v:

$$ES(T,i) = \sum\limits_{e \in q(T,i,r)} \left(\lambda_e \prod\limits_{v \in a(T,i,e)} \frac{1}{\deg_{out}(v)} \right) \tag{9.2}$$

The second derivation of TD, fair proportion or Evolutionary Distinctiveness (ED; Isaac et al. 2007), is conceptually similar to ES, but instead partitions branch lengths by the total number of species subtending it, not just the branches directly below it (see fig. 9.3C). It is calculated as:

$$ED(T,i) = \sum\limits_{e \in q(T,i,r)} \left(\lambda_e \cdot \frac{1}{S_e} \right) \tag{9.3}$$

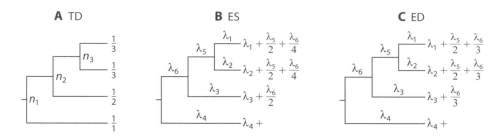

Figure 9.3. Schematic representation of how the three evolutionary distinctiveness (ED) metrics partition evolutionary information: (A) TD = Taxonomic Distinctiveness, (B) ES = Equal Splits, and (C) ED = Fair proportion or Evolutionary Distinctiveness.

where e is a branch of length λ in the set $s(T,i,r)$ connecting species i to the root r, and S_e is the number of species that descend from edge e. A nice feature of ES and ED is that they both sum to PD.

We can easily calculate ES and ED in R using the `evol.distinct()` function in *picante*. Using the Jasper Ridge data, we can calculate these two quantities by changing a single argument called `type`:

```
library(picante)
j.tree<-read.tree(".../jasper_tree.phy")
j.com<-read.csv(".../jasper_data.csv",row.names=1)
evol.distinct(j.tree,type="equal.splits")
evol.distinct(j.tree,type="fair.proportion")
```

This function returns a data frame with the specified distinctiveness values.

9.3. PRIORITIZING SPECIES BASED ON EVOLUTIONARY DISTINCTIVENESS

Above, we have considered evolutionary history in isolation; however, evolutionary history can be integrated with extinction probabilities for conservation prioritization (e.g., Witting and Loescke 1995, Faith and Walker 1996, Weitzman 1998, Redding and Mooers 2006). Isaac et al. (2007) used relative extinction risk to weight evolutionary distinctiveness, which they refer to as EDGE (Evolutionarily Distinct and Globally Endangered):

$$EDGE_i = \ln(1 + ED_i) + GE_i \cdot \ln(2) \qquad (9.4)$$

where GE is the IUCN Red List category weight (Least Concern = 0, Near Threatened and Conservation Dependent = 1, Vulnerable = 2, Endangered = 3, Critically Endangered = 4), here representing extinction risk on a log scale. Thus, EDGE values are interpreted as the log-transformed expected loss of evolutionary diversity, where each increment in Red List ranking corresponds to a doubling of extinction probability (Isaac et al. 2007). To calculate EDGE values, one must first calculate species ED (equation 9.3) and then apply equation 9.4. Here we will use the Isaac et al data available in their supplementary material on mammal ED and GE values, and calculate EDGE values:

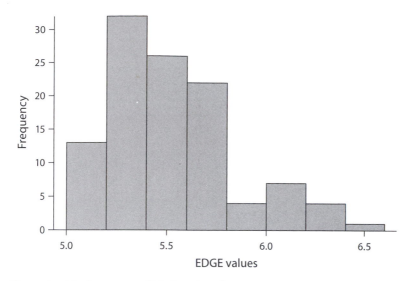

Figure 9.4. The histogram of EDGE values for mammals. From Isaac et al. (2004).

```
IP1<-read.csv("…isaac_P1.csv")
EDGE<-function(ED,GE) log(1+ED)+(GE*log(2))
vals<-EDGE(IP1$ED,IP1$GE)
hist(vals,xlab="EDGE values",col="grey",main=NULL)
```

This returns figure 9.4. For conservation purposes, it could be argued that we should prioritize those species in the right tail of this distribution—these represent those species that are most critically endangered and evolutionarily distinct.

9.3.1. Incorporating ED and Abundance

Within a defined geographical context, a species' vulnerability is a product of its representation at these scales. Species with few populations or those that occur only within a small subset of protected areas might warrant special attention. In addition, the evolutionary history encapsulated by a set of species will be differentially represented among sites (Rosauer et al. 2009, Cadotte and Davies 2010). Here we detail two approaches that allow quantification of evolutionary distinctiveness within a single species, within a multispecies site, or across larger multi-site regions.

The first metric, Rosauer's PE, developed by Rosauer et al. (2009) divides individual branch lengths by the union of the ranges of the subtending species:

$$PE(T,i) = \sum_{e \in q(T,i,r)} \frac{\lambda_e}{R_e} \tag{9.5}$$

where R_e is the clade range, or the union of the subtending species ranges.

An independently derived metric, referred to as biogeographically weighted evolutionary distinctiveness (BED; Cadotte and Davies 2010), can be calculated similarly:

$$BED(T,i) = \sum_{e \in q(T,i,r)} \frac{\lambda_e}{n_e} \tag{9.6}$$

Here n_e is the total number of populations or occupied sites that descend from branch e. The difference between BED and PE is relatively minor; BED retains all species occurrences when partitioning internal branches, while PE does not include multiple occurrences at a single site. The calculation of BED is analogous to AED (Abundance-weighted ED; see chapter 3 and Cadotte, Davies, et al. 2010), except that AED uses within-community abundances and BED uses biogeographical measures of range size and occupancy. The code to calculate BED is implicit in other metrics (e.g., H'$_{aed}$; see chapter 3), but we will use a stand-alone function that is a modified version of the one created by José Hidasi-Neto (http://rfunctions.blogspot.ca/2013/09/functions-for-phylogenetic-diversity.html). The two input variables are a phylogenetic tree and an abundance or range size vector where the elements' names are the species names. First, we need to load the function; we can either copy and paste it into the r console or we can load the function from an external directory using `source()`:

```
source("…/BED")
```

Now we will use the owl phylogeny in *ape* that we used in chapter 2, and we will create a vector of range sizes (range) that we have specified just to generate a reasonable span of values:

```
tree.owls<-read.tree(file="owl_tree.phy")
range<-c(5.5,18.7,1.2,16.3)
names(range)<-tree.owls$tip.label
BED(range,tree.owls)
```

We can visualize the BED values using the plot function with the *phylobase* package:

```
library(phylobase)
owl.bed<-phylo4d(tree.owls,tip.data=BED(range,tree.owls))
names(owl.bed@data)<-"BED"
plot(owl.bed)
```

This creates figure 9.5.
Compare these values to the ED values that are not weighed by range size, using the `evol.distinct()` function from *picante*:

```
evol.distinct(tree.owls)
```

And it is clear that ED and BED values can result in different species prioritizations (although in this example the differences are contrived). Evolutionary distinctiveness (ED) naturally returns the most distinct species, with the barn owl—*Tyto alba*—having the highest value (fig. 9.5). However, in our hypothetical region, barn owls are fairly common and despite having high ED are not highly prioritized. *Athene noctua*, or the little owl, is rare and has a moderate ED value and is thus prioritized by BED.

The sum of the species' BED values is equal to total PD, meaning that the proportion of the total PD contained within single populations or a set of sites can be simply calculated by summing BED values. Further, for species i, BED$_i$ values can be used to evaluate species relative importance values IV$_i$ within and across sites:

$$IV_i = BED_i \bigg/ \sum_{i=1}^{S} BED_i \qquad (9.7)$$

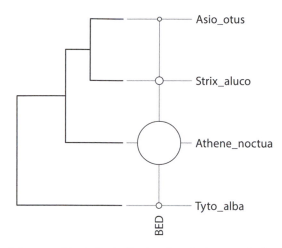

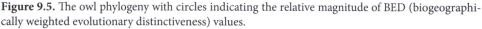

Figure 9.5. The owl phylogeny with circles indicating the relative magnitude of BED (biogeographically weighted evolutionary distinctiveness) values.

High IV species have populations, which are evolutionary distinct relative to those for populations of other species. The IV values can be summed across species occurring at a single sampling site, reserve, or at larger spatial scales. This metric assumes that the divergences among populations of a species have a length of zero, but this assumption can be modified (see appendix 2 in Cadotte, Davies, et al. 2010).

The IV values are easy to calculate in *R*:

```
BED(range,tree.owls)/sum(BED(range,tree.owls))
```

Here IV again prioritizes *Athene noctua* for conservation action.

9.4. PRIORITIZING HOTSPOTS OF EVOLUTIONARY DISTINCTIVENESS

Remember that biodiversity hotspots were assessed based on species richness and endemism (Myers et al. 2000), which are also site level measures. We can thus directly compare the sum of BED or PE values to other diversity measures (i.e., richness or abundance weighted richness) (Rosauer et al. 2009, Cadotte and Davies 2010, Tucker and Cadotte 2013). Using the BED framework, the total evolutionary distinctiveness represented by a single site *t*, with co-occurring species is:

$$ED_t = \sum_{i=1}^{S} BED_i \tag{9.8}$$

Cadotte and Davies (2010) also provide a metric to calculate the conservation value *CV* of region *L* by summing the ED_t values across sites; this is standardized by the total number of sites sampled, *N*:

$$CV_L = \sum_{t=1}^{N} \sum_{i=1}^{S} \left(\sum_{e \in q(T,i,r)} \frac{\lambda_e}{n_e} \right) \Big/ N \text{ or } CV_L = \left(\sum_{t=1}^{N} ED_t \right) \Big/ N \tag{9.9}$$

9.5. APPLYING CONSERVATION METRICS

Despite their wide appeal and clearly articulated importance, hotspots are unlikely to provide a panacea for the current biodiversity crisis, because there is no single metric that can capture all aspects of diversity that we might value. There is an amazing assortment of biodiversity metrics, and some even include subjective assessments based upon charisma, aesthetic and cultural values of species, and so forth (Ehrlich and Ehrlich 1992). Subjective measures are inherently difficult to incorporate into traditional reserve selection algorithms. We argue that metrics used for conservation prioritization ought to be as objective as possible—though we recognize that any prioritization scheme requires some subjective decisions. Even though measures that incorporate phylogenetic information appear to be suitably objective, their implementation can be problematic. One major problem is that different diversity metrics do not necessarily align, meaning that decisions need to be made about which measures to prioritize (Tucker et al. 2016). In addition, a hotspot of diversity for one taxon might not be congruent with the hotspot for another (Grenyer et al. 2006).

Compounding these difficulties, it is not always obvious that we should reduce the weight of low-diversity habitats; such places might still provide valuable ecosystem services, such as clean water and fertile soils, and be important for climate modulation. The relative gains and costs of diversity loss in different systems may also be unequal, and this might be particularly problematic if there is a nonlinear relationship between biodiversity and ecosystem services (Kareiva and Marvier 2003). For example, the loss of species from a highly diverse system might have relatively low ecosystem costs, whereas the loss of just a few species from a low-diversity or already heavily affected ecosystem could push the system toward a tipping point, where ecological costs are large and potentially irreversible. Last, we need to consider the costs of conservation action; low-diversity systems may be cheaper to protect than more species-rich environments, and perhaps our most cost-effective strategy is therefore to focus on low-to-medium diversity systems where impacts have not yet been felt.

Despite the multiple potential pitfalls and conflicts in establishing conservation priorities, it is urgent that we develop a cohesive blueprint to address the global biodiversity crisis (Mace et al. 2000). How do we begin to choose among the many different metrics that are available to us? We might take some comfort from the knowledge that alternative schemes are themselves largely complementary. Many schemes identify the same or similar areas as conservation priorities (Brooks et al. 2006); an obvious start would therefore be to focus resources on these zones of overlap. A more robust approach might be to consider aggregate metrics that consider multiple factors (including costs). We have provided one illustration of how data on rarity and phylogeny can be combined to provide a more informative measure of conservation value than either alone. We argue that maximizing the representation of phylogenetic diversity not only contributes to preserving ecosystem function (chapter 3) and the valued services they provide, but also maintains future options in an uncertain and rapidly changing world.

CHAPTER 10

Conclusion: Where To From Here?

We can learn from history on condition
that we understand it is history.
—*Roger Kaplan*

Throughout the chapters in this book we have explored the major methods and concepts in the field of ecophylogenetics. We have reviewed many of the common statistics and metrics adopted by ecologists when testing ecophylogenetic hypotheses at both small and large scales. The power of this approach is predicated on the assumption that phylogenies capture meaningful information about evolutionary influences on modern species' ecological similarities and differences (chapter 1). In this book, we have mostly used phylogenetic relationships as a given. While we discuss approaches to reconstructing phylogenetic relationships (chapter 2), this certainly deserves more critical thought: How do methodological choices about phylogenetic estimation and tree building influence our ecophylogenetic tests? For example, should we use the "best" tree in our analysis, or ought we to use, say, the 1,000 most likely trees, or sample from the Bayesian posterior distribution of trees?

Further, we haven't fully considered what these trees actually mean for our analyses. Although in chapter 5 we provide a brief overview of various evolutionary models that might describe the evolutionary trajectories of particular traits, we do not critically assess whether and how phylogenetic distances correspond to ecological differences. This link has come under increasing scrutiny in the recent literature, and ecologists need to embrace a more critical and sophisticated approach to ecophylogenetic analyses.

Beyond urging caution and critical thinking when employing phylogenetic methods, we note that there is also an underlying tension with trait-based analyses. Perhaps unsurprisingly, different researchers have differing views on interpreting what phylogeny and traits can tell us about ecological patterns. Some authors see the two as complementary, with both providing potentially meaningful information (Swenson 2014), while others believe that phylogenies simply provide "stand-ins" until the pertinent traits are uncovered. A number of analyses simply compare the explanatory power of traits versus phylogenies (e.g., Cadotte et al. 2009, Liu et al. 2015); it is then possible to select whichever explains more variation in some ecological pattern. However, given the complex relationship that may exist between phylogeny, traits, and the ecological patterns that link to them, it is not clear whether it is valid to simply compare the statistical power of the phylogeny and traits. In addition, phylogeny provides information on the evolutionary history of traits that cannot be inferred simply from community data. Such information might reveal the signature of how ecology has shaped evolution, whether through evolutionary character displacement, adaptive radiations, or niche conservatism, as some examples. Some ways forward are starting to emerge, and we revisit some of these advances in the following sections.

10.1. PREDICTING ECOLOGY FROM EVOLUTIONARY PATTERNS

Fundamental to testing many ecophylogenetic hypotheses is the translation of phylogenetic distances into expected niche differences (Letten and Cornwell 2014). To date, the community phylogenetic literature has largely focused on ecological hypotheses (e.g., competitive exclusion vs. environmental filtering) and has largely overlooked evolutionary hypotheses about trait and niche evolution, despite rapid advances in both fields. Most ecophylogenetic studies assume a positive relationship between ecological differences and phylogenetic distances (fig. 10.1A), given the premise that greater evolutionary distance has allowed species to accumulate more ecological differences. While this assumption may be common, it assumes a very particular evolutionary model, in which traits continuously diverge over time. This pattern of evolution is not commonly seen in comparative phylogenetic studies (Kelly et al. 2014), and a wide range of alternative models are available. Yet, despite extensive discussion in the comparative methods literature on models underlying trait evolution (O'Meara 2012), they are only rarely considered within ecophylogenetic studies (Peres-Neto et al. 2012, Letten and Cornwell 2014).

A number of recent reviews have stressed the importance of including explicit evolutionary models in ecophylogenetic analysis (Mouquet et al. 2012, Srivastava et al. 2012, Letten and Cornwell 2014). The simplest and most commonly employed evolutionary model is Brownian motion (BM) (Felsenstein 1985), which we reviewed in chapter 5. Under BM, trait change occurs as a random walk, with deviations sampled from a normal distribution (Felsenstein 1985, Harvey and Pagel 1991), and resulting in greater trait divergence among lineages, on average, with increasing time. Thus, the net result of BM evolution for a single trait (e.g., body size) is an increase in variance over time, but the expectation of the mean trait value does not change (fig. 10.1B).

If we assume this type of evolution, then we would expect the mean niche or trait value to remain unchanged, but that the average difference between species increases over time. Further, because we should also expect an increase in the variance of pairwise differences in species' trait values through time, BM evolution would actually result in a triangular relationship between ecological difference and phylogenetic distance (fig. 10.1C), and not a linear one as is often implicitly assumed.

Many traits reject a BM model of evolution, which gives rise to two important questions. First, does the type of evolutionary model change our expectations and how we articulate our hypotheses? Second, are there other models that might better describe patterns of trait evolution, and could such models help us to conduct more powerful tests? The answer to the first part of the second question is easy, because it is a simple yes. A detailed examination of the various evolutionary models is outside of the scope of this book; we touched on some in chapter 5, but more comprehensive discussions can be found elsewhere (Butler and King 2004, O'Meara 2012). The answer to the first question is more complicated and depends on the hypothesis of interest. If we wish to examine the relationship between phylogenetic distance and coexistence via niche differences (e.g., Slingsby and Verboom 2006, Connolly et al. 2011, Violle et al. 2011, Narwani et al. 2013, Godoy et al. 2014), then absolutely our choice of evolutionary model influences our results. However, ecophylogenetic

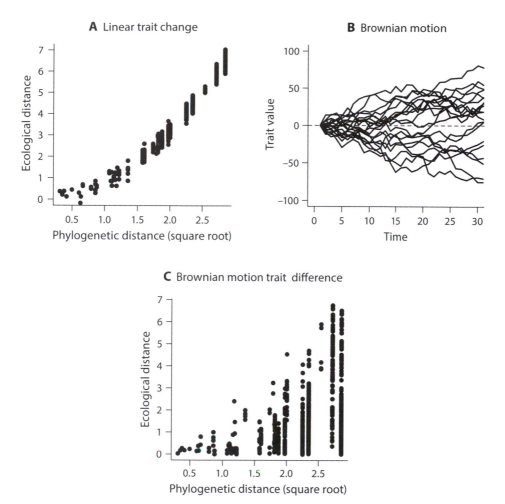

Figure 10.1. (*A*) Typical ecophylogenetic hypotheses assume that ecological differences are proportional to phylogenetic distance. (*B*) However, simple evolution models like Brownian motion predict that average trait values remain constant (*dashed lines*), but that variance increases over time, (*C*) resulting in a triangular-shaped relationship between ecological difference and phylogenetic distance.

tests that sum phylogenetic distances or test assembly using mean distances might be relatively insensitive to the particular evolutionary model assumed. Summing distances will always correspond to greater total niche differences regardless of whether the underlying trait distribution reflects increasing variance or linear divergence in absolute value. In general, we might therefore predict that assembly tests would be less sensitive to the particular model of evolution when exploring community structure in multispecies communities. For almost all evolutionary models of trait change (with the possible exception of strict punctualism), sister taxa will inherit the ancestral trait value such that very closely related species will always tend to be more ecologically similar then more distantly related species (Kelly et al. 2014), irrespective of the subsequent model of trait divergence following speciation. Therefore, we should still be able to detect evidence of limiting similarity from competitive exclusion as overdispersion, because those species that are different will always be distantly related, while closely related species will be filtered out of the assemblage (along with similar but distantly related species).

10.2. COMBINING TRAIT AND PHYLOGENETIC INFORMATION

Traits and phylogeny may be related to one another in complex ways (Ackerly 2009, Evans et al. 2009), with trait divergence, convergence, and stasis all potentially represented on a phylogeny. If traits explain more variation than phylogeny, it is not that evolution is unimportant for ecological patterns—the traits are evolutionary products. It is instead that our measure of phylogenetic relationships does not adequately capture current species differences. Additionally, different traits will undoubtedly have evolved in different ways and at different rates. Some trait differences may be well captured by a particular phylogeny, but others may not be. Perhaps surprisingly, this is also true if traits evolved following a similar underlying model, such as BM; some traits will covary closely with phylogeny, but others might show little or no phylogenetic signal at all. Whether and how to combine multiple traits into a single multivariate distance is another major question, although one we do not expand on here (but see Swenson 2014); phylogeny simply provides us with one approach. Regardless, it is always useful to ask if traits and phylogeny explain the variation in the data, if one is better than another, or if they provide complementary information. In addition, a mismatch between traits and phylogenetic expectations might itself be interesting (see chapter 6).

We suggest that in many cases phylogenies represent a more flexible tool to understand the evolution of current ecological differences than measuring a handful of traits, because the phylogenies provide not only information on current patterns, but also capture the evolutionary history that has shaped them. We now have multiple models to link evolutionary time to ecological divergence, and we can translate edge lengths into potential differences in, say, niche overlap. With traits, we often lack this insight. For example, does a 10% difference along one niche axis, such as leaf size, correspond to a 10% reduction in overall niche overlap? Because niches have many axes, the answer is likely no. But then how do we combine multiple traits into a single measure? Analyses of phylogenetic trees also have limitations. There is no direct link between phylogeny and ecological mechanism; if we are interested in herbivory resistance, for example, we may do better to measure secondary chemicals. However, it would be even better to understand the evolutionary history of these chemicals, as this might give as additional insight into coevolutionary dynamics. Further, even if traits appear to explain more variation than phylogenetic distances, phylogeny may still explain additional variation that represents unmeasured traits (Bässler et al. 2014). In addition, if we observe a correlation between a trait and some process, we would want to make sure that we have identified the correct trait by correcting for phylogenetic nonindependence in the data.

There are multiple ways that we can combine traits and phylogeny in ecophylogenetic analyses. The simplest is to generate separate distance matrices for each data type and put both distances into singular statistical models, such as multiple regressions or path analysis, to compare explanatory power and statistical interactions. However, this approach is limited because covariance between traits and phylogeny may influence analyses. Alternative approaches allow us to integrate traits and phylogeny directly. For example, we can scale edge lengths by trait values (e.g., Blomberg et al. 2003). This approach assumes the phylogenetic topology as given but allows edge lengths to reflect differences in the rates of trait evolution. Such an approach would effectively free us from the assumption that trait evolution follows a Brownian motion model, but of course requires the specification of an alternative model of how traits evolve. As we saw in chapter 5, there are many alternative models to choose from. The adoption of methods that allow us to better choose between them could enhance the ecological relevance of phylogenetic trees, although these methods have not been widely adopted in the ecophylogenetics literature.

A different approach allows us to combine phylogenetic distances (PDist) and trait or functional distances (FDist) into a single functional-phylogenetic distance (FPDist). Values for FPDist weight (a) the independent contributions of FDist and PDist and allow us to combine them nonlinearly (p), as in the following (Cadotte et al. 2013):

$$FPDist = (aPDist^p + (1-a)FDist^p)^{1/p} \tag{10.1}$$

Functional distance (FPDist) is essentially phylogenetic distances that include trait convergence and divergence; or conversely, functional distance accounting for information from unmeasured traits that have a phylogenetic signal, and thus should better estimate species functional differences. The calculation of FPDist is possible in the *pez* package using the `funct.phylo.dist()` function. This approach can be used to test assembly mechanisms or biodiversity-ecosystem relationships, or can be included in any model that employs species' distances (Cadotte et al. 2013, Bässler et al. 2014).

10.3. PHYLOGENETIC INSIGHTS INTO A CHANGING WORLD

E.O. Wilson's (1965) chapter in the seminal volume *Genetics of Colonizing Species* (edited by H. G. Baker and G. L. Stebbins) starts with a sober assessment of the current state of ecological and evolutionary theory:

> It is commonly stated that the true test of a theory is its predictive power. By this criterion we must admit to the lack of any theory worthy of the name that treats the subject of the ecology of colonization. The failure is dramatized in the case of biological control in economic entomology, a field where correct prediction is all-important to the success of costly projects.

This passage was written over half a century ago. The challenge of using theory to understand basic ecological processes and the potential impacts from invasive species and pests is as great or greater today. In addition, we now face new challenges, including climate change and extinction. We have briefly touched upon extinctions in the preceding chapters, in the following few sections we discuss how phylogeny has also helped us to develop a more predictive science of invasion biology and climate change.

10.3.1. Species Invasions

One major trend that has directly altered the composition of communities is the human-facilitated movement of species around the world. The introduction of nonindigenous species (NIS) brings together species that have been evolving separately for millions of years. When introduced, NIS bring with them a suite of traits shaped by past environments that might not match to the past of native species. As we have seen throughout this book, species' similarities and differences are fundamental in shaping ecological interactions. While taxa with shared past interactions may have diverged in key traits or behaviors allowing for their coexistence (e.g., by partitioning niche space), NIS will have had no such history with the native pool. Thus the ecological dynamics of NIS are highly unpredictable. Many NIS fail to thrive at all, while others go on to spread invasively with dramatic impacts on the ecosystems that they invade (Cadotte and Jin 2014).

Darwin (1859) first articulated the alternative expectations for the phylogenetic patterns of non-native invasions. On the one hand, he supposed that NIS distantly related to the native residents should be more successful. This is often referred to as "Darwin's naturalization hypothesis" (Daehler 2001). Elsewhere in the *Origin of the Species,* Darwin suggested that NIS should be more successful where their close relatives are found because they share the necessary adaptations for success in new environments. This is referred to as the "pre-adaptation hypothesis" (Ricciardi and Mottiar 2006). This conflict is often referred to as Darwin's "naturalization conundrum."

A number of studies representing different taxa and ecosystems have examined patterns of NIS relatedness to the communities they invade (Rejmanek and Richardson 1996, Duncan and Williams 2002, Ricciardi and Atkinson 2004, Lambdon and Hulme 2006, Strauss et al. 2006, Diez et al. 2008, Diez et al. 2009, Carboni et al. 2013, Park and Potter 2013). Some studies have found evidence that NIS are more closely related to native residents than expected by chance (e.g., Daehler 2001, Duncan and Williams 2002, Diez et al. 2008, Park and Potter 2013), while others show the opposite (e.g., Ricciardi and Atkinson 2004, Strauss et al. 2006, Strecker and Olden 2014). These mixed results hint at the need to more carefully consider the context of species invasion. Opposing findings may have examined assemblages at different spatial scales, studied different stages of invasion, used different null expectations, and considered different introduction histories (Proches et al. 2008, Thuiller et al. 2010, Cadotte 2014). For strong inference we require data on both successes and failures (introduced species that have not gone on to become invasive), as well as information across invasion stages (see e.g., Strauss et al. 2006, Bezeng et al. 2015, Li et al. 2015). Fortunately, such data are becoming more common.

10.3.2. Climate Change

Climate change represents a great unknown, and scientists are being called on to make predictions about how such change will affect global biodiversity patterns (Sala et al. 2000). There has been much work recently trying to predict how species and communities will respond to projected climate shifts (Gruter et al. 2006, Benton et al. 2007, Norf et al. 2007, Altermatt et al. 2008, Davis et al. 2010, Alexander et al. 2011, Sorte et al. 2012). The term "climate change" is a catchall that encapsulates a number of important environmental changes (e.g., temperature, precipitation, seasonality, snowpack, etc.), each of which can affect an organism's performance, phenology, and interactions with others, either in isolation or in combination (Wolkovich et al. 2012). Shifts in climate might influence the strength of the environmental filters within which community assembly mechanisms operate. Such alterations to environmental filters could result in the selection of predictable assemblages (Alexander et al. 2011), assuming that species fit closely to their climate niche. However, this ignores the importance of biotic interactions; climate change may have more pervasive and subtle impacts than just on where a species can live. Shifts in climate will also alter the timing of species activities, with inevitable effects on mutualistic, antagonistic, and competitive interactions among species (Rafferty and Ives 2012, Willmer 2012, Wolkovich et al. 2012).

Phylogenetic relationships provide information that can be used to aid in the understanding and predicting of species' range shifts (Lawing and Polly 2011) and of ultimate changes in large-scale biodiversity patterns (Thuiller et al. 2011). For example, while we may still not know exactly how thermal tolerances have evolved across a phylogeny, there is evidence that they are relatively conserved (Martinez-Meyer and Peterson 2006, Donoghue 2008, Kozak and Wiens 2010, Lawing and Polly 2011). We might, therefore, expect closely related species to respond similarly to shifts in climate. Moreover, the phylogenetic lineages present in a given region can provide information on physiological processes resulting from altered composition and environmental ecophysiological drivers. For example, information on phylogeny might help predict the presence of different photosynthetic pathways important for estimating rates of carbon sequestration at ecosystem scales (Edwards et al. 2007).

However, key questions remain. For example, as species face changing climates, what will be the relative balance between conserved traits driving range shifts and in situ evolution that allows populations to persist in suboptimal environments (Parmesan 2006)? Another emerging question is how the historical evolution of flowering time will dictate community flowering patterns with earlier springs and altered growing seasons. Harsh environments with short growing seasons appear to constrain flowering time, but in less constrained environments there is a strong phylogenetic signal to flowering time (Lessard-Therrien et al. 2014). Thus, it will be necessary to predict flowering time shifts under different environments, and phylogeny may help with this prediction (Davies et al. 2013, Mazer et al. 2013).

Finally, as community composition changes along with the timing of species interactions, the traditional assembly mechanisms structuring communities also changes. The phylogenetic methods employed throughout this book offer some insights into how we can measure these changes, and perhaps provide predictions on how these changing environments might restructure communities.

10.3.3. Phylogenetic Repercussions of Biotic Homogenization

As human activities alter habitats and abiotic conditions globally, the compositions of regional biotas are becoming more similar (McKinney and Lockwood 1999, Smith 2006, Dormann et al. 2007, McKinney and La Sorte 2007), due in part to the human-facilitated movement of species; but this increasing similarity is also caused by the range expansion of generalist species and species adapted to disturbed environments. A global analysis of long-term observations has shown that while local-scale α diversity has not systematically declined over time, the composition of the assemblages is changing in predictable ways (Dornelas et al. 2014), and this is the engine for biotic homogenization. This homogenization can be measured across a number of different metrics (Olden and Rooney 2006, Winter et al. 2009), and reveals how the multiple facets of biodiversity are converging.

As the phylogenies of regional biotas become more similar, there is a need to consider the reasons and the consequences. There may be phylogenetic reasons to expect that communities become more homogenous; for example, it could be due to the greater filtering that appears to occur in disturbed or modified environments and that selects for certain lineages (see chapter 3; Helmus et al. 2010). Nonindigenous species (NIS) are a cause of greater homogenization at larger scales because closely related NIS tend to fill out similar ranges and occur together in anthropogenic habitats (McKinney 2004, 2006; Winter et al. 2009; Cadotte, Borer, et al. 2010). The consequence of such changes is as yet unclear.

Much of this book is predicated on the premise that evolutionary history and phylogenetic relationships influence where species occur and their ecological interactions, and that this has consequences for ecosystem function and services. There is great potential in the application of these methods to address central questions on the dynamics and repercussions of biotic homogenization. What does it mean for the conservation of phylogenetic diversity? Does phylogenetic homogenization lead to homogenization of ecosystem functions and services? To date, we have barely scratched the surface in our efforts to address these questions.

10.4. WHERE TO GO FROM HERE?

Ecophylogenetics is a rapidly evolving area of research, and in five years methods might look very different than how they do today. Various core assumptions currently need to be tested and refined, and phylogenetic analyses of communities must better account for ecological, biogeographical, and evolutionary processes. Here we identify several possible trajectories for ecophylogenetic analyses in the future that we think will be important in moving the field forward.

10.4.1. Linking Coevolved Networks

Ecophylogenetic analyses most often use a single phylogeny representing a group of species from a single trophic level (e.g., a plant community). This framework assumes implicitly that trophic interactions are less important in structuring communities. However, organisms evolve in a complex network of interactions that span across trophic levels, and include mutualisms, exploitation, parasitism, and predation; and species from different trophic groups have intimately intertwined evolutionary histories. We have developed theoretical models to make predictions on the effect of one trophic level, a herbivore, on another trophic level, a plant community (Cavender-Bares et al. 2009).

In addition, empirical data suggest herbivores can significantly alter plant phylogenetic community structure (Yessoufou et al. 2013), and that the phylogenetic relatedness of plants can influence patterns of herbivory (Pearse and Hipp 2014). However, we still lack a dynamic phylogenetic framework that allows for feedback between trophic levels, and that considers community structure simultaneously across trophic components.

Advances in the modeling of interaction networks hold much promise. For example, recent models have explored the complex interconnectedness between two groups of organisms (Bascompte et al. 2003, Bastolla et al. 2009, Ings et al. 2009). A classic example of such a network would be a plant-pollinator mutualistic network (e.g., fig. 10.2). Such a network can reveal which flowers are visited by the most pollinator species and which pollinator visits the greatest number of flowers. These networks can also provide additional information that is not simple to extract from the phylogeny for a single trophic group; an example is the nestedness of interactions and the degree of specialization of these interactions (Bascompte et al. 2003). Mutualistic networks are also valuable for predicting the potential for cascading coextinctions resulting from the disappearance of mutualistic partners (Koh et al. 2004).

Figure 10.2. An example of a plant-pollinator mutualistic network. *Black squares* indicate the presence of a mutualistic association.

These types of interaction networks have often been analyzed at the species level, but the underlying phylogenies can provide a deeper understanding of species interactions and the evolution of these interactions (Rezende, Jordano, et al. 2007; Rezende, Lavabre, et al. 2007). Phylogenetic interaction networks are not in themselves new, and they have been used to provide insights into coevolutionary dynamics (Legendre et al. 2002); however, our ability to draw ecological inferences from them are developing (Rafferty and Ives 2013). Viewing a phylogenetic interaction network (e.g., host-pathogen; fig. 10.3) can reveal examples of phylogenetic concordance, where a pathogen or group of pathogens only infects a clade of closely related hosts, evolutionary generalists or specialists, or alternatively can identify cases of host switching (e.g., see the impressive work by Streicker et al. [2010] on rabies in bats). While these tools have become available, to date very little work has been done to apply them to community dynamics and assembly.

10.4.2. Linking Biogeography to Phylogenetic Analyses

As we discussed in chapter 4, the species pool used for ecophylogenetic analysis can influence the power to detect nonrandom patterns, even changing their direction and magnitude. However, there is a deeper, subtler issue: the biogeographic history of the species pool. The species pool provides the source material for local assemblages, but it is itself a product of large-scale historical and stochastic processes. For example, if we analyze local plant communities in Sweden, the presence in the species pool of a few gymnosperms, which are distantly related to all nongymnosperms, will increase the probability of detecting local phylogenetic clustering (e.g., Cadotte 2014). If we compare this to an analysis of semi-arid plant communities in western Madagascar with relatively young and small-ranged taxa, we would be more likely to detect overdispersion locally. These biogeographic processes can influence the detection of local phylogenetic patterns independent of the actual local

Plant phylogeny   Pathogenic fungi phylogeny

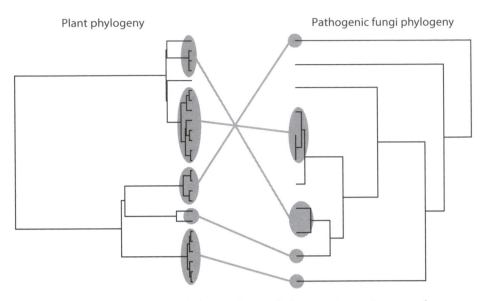

Figure 10.3. An example of a host-pathogen phylogenetic interaction network.

processes influencing community assembly. This can be both a strength and a weakness for ecophylogenetic analyses. The impact of the biogeographical pool should not be overlooked (Lessard, Borregaard, et al. 2012, Carstensen et al. 2013).

There is a further need to link predictions from evolutionary models to the biogeographical processes that have shaped the species pool. For example, an analysis of ecophylogenetic patterns among cichlid fish should explicitly consider an evolutionary model that allows for rapid recent divergence in ecological characters (e.g., Seehausen 2006). If species stemming from a recent adaptive radiation are included in an analysis with older taxa that have evolved relatively slowly, then the models used to generate hypotheses should be flexible and allow for different evolutionary rates within the phylogeny (Purvis et al. 1995).

One glaring ecological example where biogeographic context is especially important is in the analysis of native and nonindigenous communities. As we discuss above, native communities are comprised of species that have been coevolving together for much of their recent evolutionary history, and this sympatry may be responsible for driving ecological divergence, especially among closely related species (Schluter 1994). However, closely related species that are allopatric have not been subject to the same directional evolutionary pressures. Thus, when NIS are introduced into a species pool that contains close relatives, we predict ecological interactions could be either very strong, with both competing for the same resource, or very weak, with the native and non-native species having specialized on a different local resources in allopatry. In addition, it seems likely that we should model ecological differences separately for native and nonindigenous species. If NIS come from different species pools, we might assume that that they have experienced different selection pressures and competitive loads. As NIS have evolved independently from natives, we may expect NIS to appear as if they have evolved according to Brownian motion relative to the natives, which are likely to have evolved coexistence strategies with other sympatric species (Cadotte and Jin 2014). In combination, native and nonindigenous species might, therefore, be expected to show different phylogenetic community assembly patterns.

10.4.3. A Call for Carefully Designed Experiments

As we have mentioned repeatedly, the various analyses of ecophylogenetic patterns covered in this book rely on some key assumptions about the relationship between phylogenetic distance and ecological or niche differences. We have argued that more sophisticated evolutionary models need to be used. In addition, we suggest that any ecological predictions stemming from phylogenetic patterns should be tested with experiments. Ecophylogenetic experiments (Liu et al. 2011, Violle et al. 2011, Narwani et al. 2013, Godoy et al. 2014) provide perhaps the only approach for verifying whether inferences drawn from phylogenetic relationships match to ecological mechanisms. There are several potential pitfalls of which experimenters need to be careful when designing such studies. Here we outline three major issues that need to be carefully considered in any experimental tests.

The first issue involves *phylogenetic pseudoreplication*. Any good experiment requires the replication of statistically independent entities for proper inference (Hurlbert 1984). When we hear this, we naturally think of the physical design, such as randomizing the spatial arrangement of pots or aquaria. Of course treatments also need to be independent, meaning that our pots are assigned to a particular treatment independently from other pot treatments. However, an ecophylogenetic experiment that manipulates the amount of phylogenetic diversity in an assemblage might use different combinations of the same limited number of species to represent different phylogenetic diversity levels. As a result of this design, the same phylogenetic edges are present in multiple diversity treatments. The extreme example of this design is the use of a single species tested against multiple competitors of varying phylogenetic distances (fig. 10.4). In such a study, half of the smallest phylogenetic distance is completely nested within all other distance treatments. If evolution of the relevant trait(s) follows Brownian motion, then we have simply selected a single trajectory, as in figure 10.1B, and we lack an ability to estimate variance among lineages. Regardless

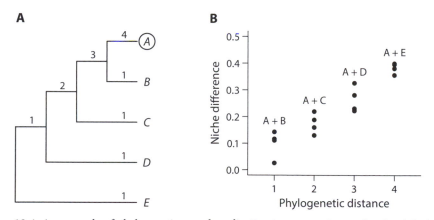

Figure 10.4. An example of phylogenetic pseudoreplication in an experiment. Species A is the test species used in all comparisons. Figure (*A*) shows the phylogeny of the species used in our hypothetical experiment, and the numbers indicate the number of times an edge is used across all treatments. Figure (*B*) shows hypothetic data, with experimental replication (*n* = 4).

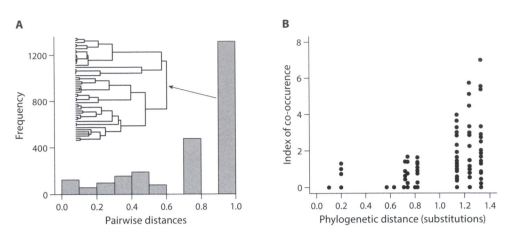

Figure 10.5. A histogram (*A*) showing the distribution of pairwise distances from a phylogeny. In a bifurcating tree, the most frequent pairwise distance will be from the root node. A sample figure (*B*) showing the natural skew in phylogenetic distances; adapted from Slingsby and Verboom (2006).

of the actual pattern observed, our inference can only be about the single evolutionary trajectory leading to our test species, and is thus not a reliable test of broader phylogenetic diversity. While the experiment in figure 10.4 is an extreme case assuming a completely pectinate phylogeny, experimenters should be wary of resampling internal edges in competition trials.

The second issue involves *unequal numbers of comparisons*. Since a phylogeny is hierarchical, the most commonly observed pairwise distances would be from the root split (fig. 10.5A). The result is that we end up with relatively few small phylogenetic distances and many large phylogenetic distances (see chapter 4). A number of recent studies exhibit this pattern. In a detailed study of the phylogenetic patterns of co-occurrence in South African sedges, Slingsby and Verboom's (2006) data exhibit this skew in phylogenetic distances (fig. 10.5B). The statistical issues associated with this are especially critical in laboratory experiments that may have limited numbers of species and thus very few short distances. Experiments should try to elevate the replication of short phylogenetic distances, at the expense of large distances if space and resources are limited (Cadotte 2013).

The third major issue involves whether *niche opportunities match nature*. Species evolve in response to many different selection pressures, and for the evolution of competitive coexistence, divergent selection could result in species utilizing different dimensions of spatial and temporal heterogeneity that reduce the strength of negative interactions. Laboratory and controlled field experiments necessarily try to remove unwanted variability in order to test specific hypotheses or mechanisms (Resetarits and Bernardo 1998). However, phylogenetic hypotheses are often ambiguous about the precise mechanisms allowing coexistence, instead preferring to frame hypotheses around a general concept of "niche differences." Controlled experiments that look for correlations between niche overlap and phylogenetic distances may actually be biased against observing a strong relationship since much of the ecologically relevant niche axes have been removed (fig. 10.6).

Experiments in controlled environments will limit the opportunity for species to take advantage of the heterogeneity they have evolved to utilize. Instead, what we should see from these experiments is a limited ability to detect niche differences and clear inequality

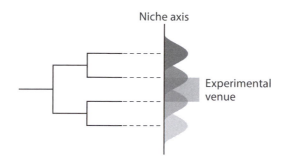

Figure 10.6. A hypothesized relationship between a phylogeny and the niche position of species. Species show clear niche segregation that is phylogenetically nonrandom. However, the experimental venue constrains the niche dimension, which reduces the likelihood of observing a correlation between niche differences and phylogenetic differences.

in species' fitnesses, because a subset of species may be well suited to the environment represented by the experimental venue. Thus, to fully assess the evolution of niche differences, experiments need to be carefully matched to species' resource and habitat requirements, and ecologically relevant heterogeneity needs to be included.

10.5. HERALDING THE ECOLOGY-EVOLUTION SYNTHESIS?

It is well documented that the studies of ecology and evolution were estranged for much of the twentieth century (Kohler 2002). However, with the advancement of population genetics and the ability to analyze transcription products within organisms, population ecology has served as the vanguard for the (re)merging of ecology and evolution (Johnson and Stinchcombe 2007). This is because population growth and individual fitness are intimately tied to competition among genotypes resulting in changes in gene frequencies (Sakai 1965). These types of ecoevolutionary dynamics operate over short time scales and explain dynamics within a population or among interacting species (i.e., community genetics). The field of ecophylogenetics emphasizes how the evolutionary past informs where species can occur and how they interact with one another today. In this framework, it is evolutionary history that informs population fitness.

We are in the midst of a new synthesis of ecology and evolution, driven by advances in population genetics and ecophylogenetics. But there remains a risk that a new divide might emerge to reinforce a border between within-species population research and work done on communities and ecosystems. This new synthesis must therefore be able to embrace theories, concepts, and tools that allow us to bridge short-term dynamics with the patterns generated through deep evolutionary time. Much of this book discusses patterns at the species level and higher, but many methods can be applied equally across multiple levels of organization, from species to populations, colonies, individuals, and so forth. Along with some of the advances we have discussed in this book, we therefore suggest future work should also aim to merge within-species or population adaptive processes with among-species ecological divergence. There are several possibilities that could be explored. For example, different models can inform different parts of a phylogeny (e.g., population genetics coalescence models below the species model, and ancestral state reconstructions above). In addition, reticulate networks that capture gene flow between populations and species could replace strictly hierarchical phylogenetic trees. Many (but not all) phylogenetic metrics can be applied to networks just as easily as they can be applied to phylogenetic trees. It would be possible to also incorporate different gene trees representing different evolutionary histories by, for example, using mitochondrial and nuclear DNA and coding and noncoding regions, or targeted genes with particular functions or under strong selection. As the phylogenetics revolution that emerged in the 1990s comes together with the current revolution in genomics, we suggest there is much potential for new insights and scope for growth.

The recent merger of evolution and ecology has led to exciting new hypotheses and methods, and has stimulated a lot of recent research. We hope that this excitement continues into the future and serves as a harbinger of a broader synthesis of ecology and evolution.

GLOSSARY

◇◇◇◇◇◇◇◇◇◇◇◇◇◇◇◇◇◇◇

Abiotic interactions: Organismal interactions with the physical environment. Includes climatic conditions, resource availability, substrates, and so forth.

Alpha diversity: Local, within-habitat diversity.

Autapomorphy: A derived trait unique to a specific terminal lineage.

Beta diversity: Among-habitat diversity or turnover. Usually measured as the average amount of diversity not found in local communities, or else as an average dissimilarity among communities.

Bifurcating tree: A phylogenetic tree in which all nodes split into two descending lineages.

Biodiversity: A catchall term meaning the diversity of life; can include multiple levels of biological organization (e.g., genetic variability, number of species, number of ecosystems, etc.).

Biodiversity facets: Different measures or aspects of biodiversity. For example, species richness, phylogenetic diversity, and functional diversity are different facets.

Biotic interactions: Organismal interactions with other organisms. Includes a finite number of possible interaction types (e.g., competition, predation, facilitation, etc.).

Brownian motion: An evolutionary model in which descendent species diverge from an ancestor by accumulating random differences in a manner analogous to a random walk.

Character displacement: Differences among similar or closely related species that are increased where the species co-occur.

Continuous traits: Traits that are measured and take on any value (e.g., height, mass, etc.).

Convergent evolution: see Homoplasy

Clustering: see Underdispersion

Dependent variable: The variable to be predicted or explained in an analysis; usually the subject of a hypothesis.

Directional selection: Increased fitness of phenotypes toward one end of a distribution of phenotypes.

Functional diversity: Trait-based measures of diversity. High values correspond to greater average trait differences than lower values.

Gamma diversity: Total regional diversity.

Homoplasy: Phenotypes shared by two species that are not present in their ancestors. Also called convergent evolution.

Independent variable: The variable used to explain variation in a variable of interest.

Latitudinal diversity gradient: The observation that biodiversity generally declines with increasing latitude.

Macroevolution: Evolutionary patterns or change described at or above the species level.

Multivariate analysis: Statistical analyses that use multiple measured variables.

Natural selection: The process by which some traits are favored over others, and the relative frequency of favored traits in a population increases over time.

Niche: An organism's place or role in nature. Modern theory often defines a niche according to an organism's resource requirements.

Niche conservatism: The tendency for species to retain ancestral phenotypes.

Niche filling: The idea that niche space is finite and can support a limited number of species.

Niche modeling: see Species distribution modeling.

Null model/distribution: A process or randomization that generates a statistic or community expectations without hypothesized structuring mechanisms.

Overdispersion: When the phylogenetic distances that separate co-occurring species are larger than expected by chance.

Patristic Distance: The distance between two tips measured on a phylogenetic tree.

Phylogenetic signal: A test of the propensity for closely related species to possess more similar trait values than two distantly related species. Often compared to the expectation from a Brownian motion model of trait evolution.

Phenology: The timing of regular or seasonally timed organismal events (e.g., flowering).

Plasticity: Change in organismal phenotype, often during development, in response to environmental conditions.

Polytomy: A node in a phylogenetic tree that gives rise to more than two descendants.

Random Walk: A path that consists of a series of steps taken at random.

Speciation: The evolutionary processes that result in new species formation. These processes occur as a result of genetic, geographic, and behavioral factors that separate formerly interbreeding populations.

Species distribution modeling: Statistical routines that use occurrence and environmental data to predict where species could occur.

Species pool: A species list representing some larger scale from which species in local assemblages are drawn.

Synapomorphy: A trait shared by two or more species because it was present in their common ancestor.

Ultrametric tree: A tree in which all terminal branches have the same tip-to-root distance.

Underdispersion: When the phylogenetic distances that separate co-occurring species are smaller than expected by chance.

Variance-covariance matrix: A square matrix in which the off-diagonal elements are the total shared branch lengths between two species, and the diagonal elements are the unshared branch lengths.

Yule tree/model: A speciation model based on a per lineage exponential species birth rate. Often referred to as a pure-birth model because it does not model extinctions.

REFERENCES

Abouheif, E. 1999. A method for testing the assumption of phylogenetic independence in comparative data. *Evolutionary Ecology Research* 1:895–909.

Abrams, P. 1983. The theory of limiting similarity. *Annual Review of Ecology and Systematics* 14:359–376.

Ackerly, D. 2009. Conservatism and diversification of plant functional traits: Evolutionary rates versus phylogenetic signal. *Proceedings of the National Academy of Sciences of the United States of America* 106:19699–19706.

Ackerly, D., D. Schwilk, and C. Webb. 2006. Niche evolution and adaptive radiation: Testing the order of trait divergence. *Ecology* 87:S50–S61.

Ackerly, D. D., and M. J. Donoghue. 1995. Phylogeny and ecology reconsidered. *Journal of Ecology* 83:730–733.

Adler, P. B., J. HilleRisLambers, and J. M. Levine. 2007. A niche for neutrality. *Ecology Letters* 10:95–104.

Alatalo, R. V. 1982. Bird species distribution in the Galapagos and other archipelagoes: Competition or chance? *Ecology* 63:881–887.

Aldous, D. J. 2001. Stochastic models and descriptive statistics for phylogenetic trees, from Yule to today. *Statistical Science* 16:23–34.

Alexander, J. M., C. Kueffer, C. C. Daehler, P. J. Edwards, A. Pauchard, T. Seipel, and M. Consortium. 2011. Assembly of nonnative floras along elevational gradients explained by directional ecological filtering. *Proceedings of the National Academy of Sciences of the United States of America* 108:656–661.

Allen, B., M. Kon, and Y. Bar-Yam. 2009. A new phylogenetic diversity measure generalizing the Shannon index and its application to Phyllostomid bats. *American Naturalist* 174:236–243.

Alroy, J. 1999. The fossil record of North American mammals: Evidence for a Paleocene evolutionary radiation. *Systematic Biology* 48:107–118.

Altermatt, F., V. I. Pajunen, and D. Ebert. 2008. Climate change effects colonization dynamics in a metacommunity of three Daphnia species. *Global Change Biology* 14:1209–1220.

Anderson, M. J., T. O. Crist, J. M. Chase, M. Vellend, B. D. Inouye, A. L. Freestone, N. J. Sanders, et al. 2011. Navigating the multiple meanings of β diversity: A roadmap for the practicing ecologist. *Ecology Letters* 14:19–28.

Andrewartha, H. G., and C. Birch. 1954. *The Distribution and Abundance of Animals*. University of Chicago Press, Chicago, IL, US.

APG (Angiosperm Phylogeny Group). 1998. An ordinal classification for the families of flowering plants. *Annals of the Missouri Botanical Garden* 85:531–553.

Archibald, J. K., M. E. Mort, and D. J. Crawford. 2003. Bayesian inference of phylogeny: A nontechnical primer. *Taxon* 52:187–191.

Armbruster, W. S. 1992. Phylogeny and the evolution of plant-animal interactions. *Bioscience* 42:12–20.

Arnold, C., L. J. Matthews, and C. L. Nunn. 2010. The 10kTrees website: A new online resource for primate phylogeny. *Evolutionary Anthropology: Issues, News, and Reviews* 19:114–118.

Arrhenius, O. 1921. Species and area. *Journal of Ecology* 9:95–99.

Avise, J. C. 2000. *Phylogeography: The History and Formation of Species*. Harvard University Press, Cambridge, MA.

Baker, H. G. and G.L. Stebbins, eds. 1965. *The Genetics of Colonizing Species*. Academic Press, New York, NY.

Balvanera, P., A. B. Pfisterer, N. Buchmann, J. S. He, T. Nakashizuka, D. Raffaelli, and B. Schmid. 2006. Quantifying the evidence for biodiversity effects on ecosystem functioning and services. *Ecology Letters* 9:1146–1156.

Barker, G. M. 2002. Phylogenetic diversity: A quantitative framework for measurement of priority and achievement in biodiversity conservation. *Biological Journal of the Linnean Society* 76:165–194.

Barnagaud, J.-Y., W. D. Kissling, B. Sandel, W. L. Eiserhardt, C. H. Sekercioglu, B. J. Enquist, C. Tsirogiannis, and J.-C. Svenning. 2014. Ecological traits influence the phylogenetic structure of bird species co-occurrences worldwide. *Ecology Letters* 17:811–820.

Barraclough, T. G., S. Nee, and P. H. Harvey. 1998. Comment: Sister-group analysis in identifying correlates of diversification. *Evolutionary Ecology* 12:751–754.

Barraclough, T. G., and A. P. Vogler. 2000. Detecting the geographic pattern of speciation from species-level phylogenies. *American Naturalist* 155:419–434.

Barraclough, T. G., A. P. Vogler, and P. H. Harvey. 1998. Revealing the factors that promote speciation. *Philosophical Transactions of the Royal Society of London B, Biological Sciences* 353:241–249.

Bascompte, J., P. Jordano, C. J. Melián, and J. M. Olesen. 2003. The nested assembly of plant–animal mutualistic networks. *Proceedings of the National Academy of Sciences of the United States of America* 100:9383–9387.

Bässler, C., R. Ernst, M. Cadotte, C. Heibl, and J. Müller. 2014. Near-to-nature logging influences fungal community assembly processes in a temperate forest. *Journal of Applied Ecology* 51:939–948.

Bastolla, U., M. A. Fortuna, A. Pascual-Garcia, A. Ferrera, B. Luque, and J. Bascompte. 2009. The architecture of mutualistic networks minimizes competition and increases biodiversity. *Nature* 458:1018–1020.

Belyea, L. R., and J. Lancaster. 1999. Assembly rules within a contingent ecology. *Oikos* 86:402–416.

Bennett, J. A., E. G. Lamb, J. C. Hall, W. M. Cardinal-McTeague, and J. F. Cahill. 2013. Increased competition does not lead to increased phylogenetic overdispersion in a native grassland. *Ecology Letters* 16:1168–1176.

Benton, T. G., M. Solan, J. M. J. Travis, and S. M. Sait. 2007. Microcosm experiments can inform global ecological problems. *Trends in Ecology & Evolution* 22:516–521.

Berlow, E. L. 1997. From canalization to contingency: Historical effects in a successional rocky intertidal community. *Ecological Monographs* 67:435–460.

Bezeng, S. B., T. J. Davies, K. Yessoufou, O. Maurin, and M. Van der Bank. 2015. Revisiting Darwin's naturalization conundrum: Explaining invasion success of non-native trees and shrubs in southern Africa. *Journal of Ecology* 103:871–879.

Bininda-Emonds, O. R., J. L. Gittleman, and M. A. Steel. 2002. The (super) tree of life: Procedures, problems, and prospects. *Annual Review of Ecology and Systematics* 33:265–289.

Bininda-Emonds, O.R.P., M. Cardillo, K. E. Jones, R.D.E. MacPhee, R.M.D. Beck, R. Grenyer, S. A. Price, et al. 2007. The delayed rise of present-day mammals. *Nature* 446:507–512.

Bino, G., D. Ramp, and R. T. Kingsford. 2013. Niche evolution in Australian terrestrial mammals? Clarifying scale-dependencies in phylogenetic and functional drivers of co-occurrence. *Evolutionary Ecology* 27:1159–1173.

Bivand, R. S., E. J. Pebesma, and V. Gómez-Rubio. 2008. *Applied Spatial Data Analysis with R*. Springer, New York, NY.

Blackburn, T. M., and K. J. Gaston. 2006. There's more to macroecology than meets the eye. *Global Ecology and Biogeography* 15:537–540.

Blomberg, S. P., T. Garland, and A. R. Ives. 2003. Testing for phylogenetic signal in comparative data: Behavioral traits are more labile. *Evolution* 57:717–745.

Blossey, B., and R. Notzold. 1995. Evolution of increased competitive ability in invasive nonindigenous plants: A hypothesis. *Journal of Ecology* 83:887–889.

Blytt, A. 1886. On variations of the climate in the course of time. *Nature* 34:239–242.

Boettiger, C., and D. Temple Lang. 2012. Treebase: An R package for discovery, access and manipulation of online phylogenies. *Methods in Ecology and Evolution* 3:1060–1066.

Böhning-Gaese, K., and R. Oberrath. 1999. Phylogenetic effects on morphological, life-history, behavioural and ecological traits of birds. *Evolutionary Ecology Research* 1999:347–364.

Bollback, J. P. 2006. SIMMAP: Stochastic character mapping of discrete traits on phylogenies. *BMC Bioinformatics* 7:88.

Bolnick, D. I. 2004. Waiting for sympatric speciation. *Evolution* 58:895–899.

Bray, J. R., and J. T. Curtis. 1957. An ordination of the upland forest communities of southern Wisconsin. *Ecological Monographs* 27:326–349.

Britton, T., B. Oxelman, A. Vinnersten, and K. Bremer. 2002. Phylogenetic dating with confidence intervals using mean path lengths. *Molecular Phylogenetics and Evolution* 24:58–65.

Brooks, C. P. 2006. Quantifying population substructure: Extending the graph-theoretic approach. *Ecology* 87:864–872.

Brooks, T. M., R. A. Mittermeier, G.A.B. da Fonseca, J. Gerlach, M. Hoffmann, J. F. Lamoreux, C. G. Mittermeier,et al. 2006. Global biodiversity conservation priorities. *Science* 313:58–61.

Brown, W. L., and E. O. Wilson. 1956. Character displacement. *Systematic Zoology* 5:49–64.

Brunbjerg, A. K., F. Borchsenius, W. L. Eiserhardt, R. Ejrnaes, and J.-C. Svenning. 2012. Disturbance drives phylogenetic community structure in coastal dune vegetation. *Journal of Vegetation Science* 23:1082–1094.

Brunbjerg, A. K., J. Cavender-Bares, W. L. Eiserhardt, R. Ejrnaes, L. W. Aarssen, H. L. Buckley, E. Forey, et al. 2014. Multi-scale phylogenetic structure in coastal dune plant communities across the globe. *Journal of Plant Ecology* 7:101–114.

Bryant, J. A., C. Lamanna, H. Morlon, A. J. Kerkhoff, B. J. Enquist, and J. L. Green. 2008. Microbes on mountainsides: Contrasting elevational patterns of bacterial and plant diversity. *Proceedings of the National Academy of Sciences of the United States of America* 105:11505–11511.

Burnham, K. P., and D. R. Anderson. 2002. *Model Selection and Multimodel Inference: A Practical Information-Theoretic Approach*. Springer Science and Business Media, New York, NY.

Butler, M. A., and A. A. King. 2004. Phylogenetic comparative analysis: A modeling approach for adaptive evolution. *American Naturalist* 164:683–695.

Cadotte, M. W. 2006. Metacommunity influences on community richness at multiple spatial scales: A microcosm experiment. *Ecology* 87:1008–1016.

Cadotte, M. W. 2007. Concurrent niche and neutral processes in the competition-colonization model of species coexistence. *Proceedings of the Royal Society of London B, Biological Sciences* 274:2739–2744.

Cadotte, M. W. 2013. Experimental evidence that evolutionarily diverse assemblages result in higher productivity. *Proceedings of the National Academy of Sciences of the United States of America* 110:8996–9000.

Cadotte, M. W. 2014. Including distantly related taxa can bias phylogenetic tests. *Proceedings of the National Academy of Sciences of the United States of America* 111:E536.

Cadotte, M. W. 2015. Phylogenetic diversity-ecosystem function relationships are insensitive to phylogenetic edge lengths. *Functional Ecology* 29:718–723.

Cadotte, M. W., C. H. Albert, and S. C. Walker. 2013. The ecology of differences: Integrating evolutionary and functional distances. *Ecology Letters* 16:1234–1244.

Cadotte, M. W., E. T. Borer, E. W. Seabloom, J. Cavender-Bares, W. S. Harpole, E. Cleland, and K. F. Davies. 2010. Phylogenetic patterns differ for native and exotic plant communities across a richness gradient in Northern California. *Diversity and Distributions* 16:892–901.

Cadotte, M. W., B. J. Cardinale, and T. H. Oakley. 2008. Evolutionary history and the effect of biodiversity on plant productivity. *Proceedings of the National Academy of Sciences of the United States of America* 105:17012–17017.

Cadotte, M. W., K. Carscadden, and N. Mirotchnick. 2011. Beyond species: Functional diversity and the maintenance of ecological processes and services. *Journal of Applied Ecology* 48:1079–1087.

Cadotte, M. W., J. Cavender-Bares, D. Tilman, and T. H. Oakley. 2009. Using phylogenetic, functional and trait diversity to understand patterns of plant community productivity. *PLoS ONE* 4:e5695.

Cadotte, M. W., and T. J. Davies. 2010. Rarest of the rare: Advances in combining evolutionary distinctiveness and scarcity to inform conservation at biogeographical scales. *Diversity and Distributions* 16:376–385.

Cadotte, M. W., T. J. Davies, J. Regetz, S. W. Kembel, E. E. Cleland, and T. H. Oakley. 2010. Phylogenetic diversity metrics for ecological communities: Integrating species richness, abundance and evolutionary history. *Ecology Letters* 13:96–105.

Cadotte, M. W., R. Dinnage, and D. Tilman. 2012. Phylogenetic diversity promotes ecosystem stability. *Ecology* 93:s223–233.

Cadotte, M. W., T. Fukami, and S. M. McMahon. 2006. *Conceptual Ecology and Invasion Biology: Reciprocal Approaches to Nature*. Springer, Dordrecht, The Netherlands.

Cadotte, M. W., M. A. Hamilton, and B. R. Murray. 2009. Phylogenetic relatedness and plant invader success across two spatial scales. *Diversity and Distributions* 15:481–488.

Cadotte, M. W., and L. S. Jin. 2014. All in the family: Relatedness and the success of introduced species, in *Invasive Species in a Globalized World*, ed. R. P. Keller, M. W. Cadotte, and G. Sandiford, pp. 147–162. University of Chicago Press, Chicago, IL, US.

Carboni, M., T. Muenkemueller, L. Gallien, S. Lavergne, A. Acosta, and W. Thuiller. 2013. Darwin's naturalization hypothesis: Scale matters in coastal plant communities. *Ecography* 36:560–568.

Cardillo, M., J. L. Gittleman, and A. Purvis. 2008. Global patterns in the phylogenetic structure of island mammal assemblages. *Proceedings of the Royal Society of London B, Biological Sciences* 275:1549–1556.

Cardillo, M., G. M. Mace, J. L. Gittleman, and A. Purvis. 2006. Latent extinction risk and the future battlegrounds of mammal conservation. *Proceedings of the National Academy of Sciences of the United States of America* 103:4157–4161.

Cardillo, M., G. M. Mace, K. E. Jones, J. Bielby, O.R.P. Bininda-Emonds, W. Sechrest, C.D.L. Orme, and A. Purvis. 2005. Multiple causes of high extinction risk in large mammal species. *Science* 309:1239–1241.

Cardinale, B. J., D. S. Srivastava, J. E. Duffy, J. P. Wright, A. L. Downing, M. Sankaran, and C. Jouseau. 2006. Effects of biodiversity on the functioning of trophic groups and ecosystems. *Nature* 443:989–992.

Carstensen, D. W., J. P. Lessard, B. G. Holt, M. Krabbe Borregaard, and C. Rahbek. 2013. Introducing the biogeographic species pool. *Ecography* 36:1310–1318.

Case, T. J. 1990. Invasion resistance arises in strongly interacting species-rich model competition communities. *Proceedings of the National Academy of Sciences of the United States of America* 87:9610–9614.

Cavender-Bares, J., A. Keen, and B. Miles. 2006. Phylogenetic structure of Floridian plant communities depends on taxonomic and spatial scale. *Ecology* 87:S109–S122.

Cavender-Bares, J., K. H. Kozak, P.V.A. Fine, and S. W. Kembel. 2009. The merging of community ecology and phylogenetic biology. *Ecology Letters* 12:693–715.

Cavender-Bares, J., and A. Wilczek. 2003. Integrating micro- and macroevolutionary processes in community ecology. *Ecology* 84:592–597.

Chao, A., C. H. Chiu, and L. Jost. 2010. Phylogenetic diversity measures based on Hill numbers. *Philosophical Transactions of the Royal Society of London B, Biological Sciences* 365:3599–3609.

Chase, J. M., and M. A. Leibold. 2002. Spatial scale dictates the productivity-biodiversity relationship. *Nature* 416:427–430.

Chave, J. 2004. Neutral theory and community ecology. *Ecology Letters* 7:241–253.

Chesser, R. T., and R. M. Zink. 1994. Modes of speciation in birds: A test of Lynch's method. *Evolution* 48:490–497.

Chesson, P. 2000. Mechanisms of maintenance of species diversity. *Annual Review of Ecology and Systematics* 31:343–366.

Clark, J. S. 2009. Beyond neutral science. *Trends in Ecology & Evolution* 24:8–15.

Clarke, K. R., and R. M. Warwick. 1998. A taxonomic distinctness index and its statistical properties. *Journal of Applied Ecology* 35:523–531.

Clarke, K. R., and R. M. Warwick. 2001. A further biodiversity index applicable to species lists: Variation in taxonomic distinctness. *Marine Ecology-Progress Series* 216:265–278.

Clements, F. E. 1928. *Plant Succession and Indicators*. H. W. Wilson, New York, NY.

Colwell, R. K., and G. C. Hurt. 1994. Nonbiological gradients in species richness and a spurious Rapoport effect. *American Naturalist* 144:570–595.

Connell, J. H. 1980. Diversity and the coevolution of competitors, or the ghost of competition past. *Oikos* 35:131–138.

Connolly, J., M. W. Cadotte, C. Brophy, Å. Dooley, J. Finn, L. Kirwan, C. Roscher, and A. Weigelt. 2011. Phylogenetically diverse grasslands are associated with pairwise interspecific processes that increase biomass. *Ecology* 92:1385–1392.

Connor, E. F., and E. D. McCoy. 1979. The statistics and biology of the species-area relationship. *American Naturalist* 113:791–833.

Connor, E. F., and D. Simberloff. 1979. The assembly of species communities: Chance or competition? *Ecology* 60:1132–1140.

Connor, E. F., and D. Simberloff. 1984. Neutral models of species co-occurrence patterns, in *Ecological Communities: Conceptual Issues and the Evidence*, ed. D. R. Strong, D. Simberloff, L. G. Abele, and A. Thistle, pp. 341–343. Princeton University Press, Princeton, NJ, US.

Connor, E. F., and D. Simberloff. 1986. Competition, scientific method, and null models in ecology. *American Scientist* 74:155–162.

Cooper, N., and A. Purvis. 2010. Body size evolution in mammals: Complexity in tempo and mode. *American Naturalist* 175:727–738.

Cooper, N., J. Rodríguez, and A. Purvis. 2008. A common tendency for phylogenetic overdispersion in mammalian assemblages. *Proceedings of the Royal Society of London B, Biological Sciences* 275:2031–2037.

Corey, S. J., and T. A. Waite. 2008. Phylogenetic autocorrelation of extinction threat in globally imperilled amphibians. *Diversity and Distributions* 14:614–629.

Cornwell, W. K., and D. D. Ackerly. 2009. Community assembly and shifts in plant trait distributions across an environmental gradient in coastal California. *Ecological Monographs* 79:109–126.

Cottenie, K. 2005. Integrating environmental and spatial processes in ecological community dynamics. *Ecology Letters* 8:1175–1182.

Cowles, H. C. 1911. The causes of vegetative cycles. *Botanical Gazette* 51:161–183.

Coyne, A. C., and H. A. Orr. 2004. *Speciation*. Sinaur, Sunderland, MA, US.

Crawley, M. J., and J. E. Harral. 2001. Scale dependence in plant biodiversity. *Science* 291:864–868.

Cronquist, A. 1968. *The Evolution and Classification of Flowering Plants*. Houghton Mifflin, Boston, MA, US.

Crozier, R. H. 1997. Preserving the information content of species: Genetic diversity, phylogeny, and conservation worth. *Annual Review of Ecology and Systematics* 28:243–268.

Crozier, R. H., L. J. Dunnett, and P. M. Agapow. 2005. Phylogenetic biodiversity assessment based on systematic nomenclature. *Evolutionary Bioinformatics* 1:11–36.

Cunningham, C., H. Zhu, and D. Hillis. 1998. Best-fit maximum-likelihood models for phylogenetic inference: Empirical tests with known phylogenies. *Evolution* 52:978–987.

D'Antonio, C. M., and P. M. Vitousek. 1992. Biological invasions by exotic grasses, the grass fire cycle, and global change. *Annual Review of Ecology and Systematics* 23:63–87.

Daehler, C. C. 2001. Darwin's naturalization hypothesis revisited. *American Naturalist* 158:324–330.

Darwin, C. 1859. *On the Origin of Species by Means of Natural Selection*. John Murray, London.

Davies, T. J., A. P. Allen, L. Borda-de-Água, J. Regetz, and C. J. Melián. 2011. Neutral biodiversity theory can explain the imbalance of phylogenetic trees but not the tempo of their diversification. *Evolution* 65:1841–1850.

Davies, T. J., T. G. Barraclough, M. W. Chase, P. S. Soltis, D. E. Soltis, and V. Savolainen. 2004. Darwin's abominable mystery: Insights from a supertree of the angiosperms. *Proceedings of the National Academy of Science of the United States of America* 101:1904–1909.

Davies, T. J., and L. B. Buckley. 2012. Exploring the phylogenetic history of mammal species richness. *Global Ecology and Biogeography* 21:1096–1105.

Davies, T. J., and M. W. Cadotte. 2011. Quantifying biodiversity: Does it matter what we measure? in *Biodiversity Hotspots*, ed. F. E. Zachos, and J. C. Habel, pp. 43–60. Springer, Heidelberg, Germany.

Davies, T. J., N. Cooper, J.A.F. Diniz, G. H. Thomas, and S. Meiri. 2012. Using phylogenetic trees to test for character displacement: A model and an example from a desert mammal community. *Ecology* 93:S44–S51.

Davies, T. J., S. A. Fritz, R. Grenyer, C.D.L. Orme, J. Bielby, O.R.P. Bininda-Emonds, M. Cardillo, et al. 2008. Phylogenetic trees and the future of mammalian biodiversity. *Proceedings of the National Academy of Sciences of the United States of America* 105:11556–11563.

Davies, T. J., N.J.B. Kraft, N. Salamin, and E. M. Wolkovitch. 2012. Incompletely resolved phylogenetic trees inflate estimates of phylogenetic conservatism. *Ecology* 93:242–247.

Davies, T. J., S. Meiri, T. G. Barraclough, and J. L. Gittleman. 2007. Species co-existence and character divergence across carnivores. *Ecology Letters* 10:146–152.

Davies, T. J., E. M. Wolkovich, N. J. Kraft, N. Salamin, J. M. Allen, T. R. Ault, J. L. Betancourt, et al. 2013. Phylogenetic conservatism in plant phenology. *Journal of Ecology* 101:1520–1530.

Davis, C. C., C. G. Willis, R. B. Primack, and A. J. Miller-Rushing. 2010. The importance of phylogeny to the study of phenological response to global climate change. *Philosophical Transactions of the Royal Society of London B, Biological Sciences* 365:3201–3213.

Day, T. A., and K. A. Young. 2004. Competitive and facilitative evolutionary diversification. *BioScience* 54:101–109.

Dayan, T., and D. Simberloff. 1994. Character displacement, sexual size dimorphism, and morphological variation among British and Irish mustelids. *Ecology* 75:1063–1073.

Dayan, T., and D. Simberloff. 1998. Size patterns among competitors: Ecological character displacement and character release in mammals, with special reference to island populations. *Mammal Review* 28:99–124.

Dayan, T., D. Simberloff, E. Tchernov, and Y. Yom-Tov. 1989. Inter- and intraspecific character displacement in mustelids. *Ecology* 70:1526–1539.

Dayan, T., D. Simberloff, E. Tchernov, and Y. Yom-Tov. 1990. Feline canines: Community-wide character displacement in the small cats of Israel. *American Naturalist* 136:39–60.

Dayan, T., D. Simberloff, E. Tchernov, and Y. Yom-Tov. 1992. Canine carnassials: Character displacement among the wolves, jackals, and foxes of Israel. *Biological Journal of the Linnean Society* 45:315–331.

De Bello, F. 2012. The quest for trait convergence and divergence in community assembly: Are null-models the magic wand? *Global Ecology and Biogeography* 21:312–317.

De Bello, F., M. Vandewalle, T. Reitalu, J. Leps, H. C. Prentice, S. Lavorel, and M. T. Sykes. 2013. Evidence for scale- and disturbance-dependent trait assembly patterns in dry semi-natural grasslands. *Journal of Ecology* 101:1237–1244.

De Oliveira, T. G., M. A. Torato, L. Silveira, D. B. Kasper, F. D. Mazim, M. Lucherini, A. Jácomo, et al. 2010. Ocelot ecology and its effect on the small-felid guild in the lowland neotropics, in *Biology and Conservation of Wild Felids*, ed. D. W. Macdonald and A. J. Loveridge, pp. 559–580. Oxford University Press, New York, NY.

Diamond, J. 1986. Evolution of ecological segregation in the New Guinea montane avifauna, in *Community Ecology*, ed. J. Diamond and T. J. Case, pp. 98–125. Harper and Row, New York, NY.

Diamond, J. M. 1975. Assembly of species communities, in *Ecology and Evolution of Communities*, ed. M. L. Cody and J. M. Diamond, pp. 343–444. Harvard University Press, Cambridge, MA, US.

Diaz, S., and M. Cabido. 2001. Vive la difference: Plant functional diversity matters to ecosystem processes. *Trends in Ecology & Evolution* 16:646–655.

Diekmann, U., and M. Doebeli. 1999. On the origin of species by sympatric speciation. *Nature* 400:354–357.

Diez, J. M., J. J. Sullivan, P. E. Hulme, G. Edwards, and R. P. Duncan. 2008. Darwin's naturalization conundrum: Dissecting taxonomic patterns of species invasions. *Ecology Letters* 11:674–681.

Diez, J. M., P. A. Williams, R. P. Randall, J. J. Sullivan, P. E. Hulme, and R. P. Duncan. 2009. Learning from failures: Testing broad taxonomic hypotheses about plant naturalization. *Ecology Letters* 12:1174–1183.

Dobson, G. P., and J. P. Headrick. 1995. Bioenergetic scaling: Metabolic design and body-size constraints in mammals. *Proceedings of the National Academy of Sciences of the United States of America* 92:7317–7321.

Donoghue, M. J. 2008. A phylogenetic perspective on the distribution of plant diversity. *Proceedings of the National Academy of Sciences of the United States of America* 105:11549–11555.

Dormann, C. F., O. Schweiger, I. Augenstein, D. Bailey, R. Billeter, G. de Blust, R. DeFilippi, et al. 2007. Effects of landscape structure and land-use intensity on similarity of plant and animal communities. *Global Ecology and Biogeography* 16:774–787.

Dornelas, M., N. J. Gotelli, B. McGill, H. Shimadzu, F. Moyes, C. Sievers, and A. E. Magurran. 2014. Assemblage time series reveal biodiversity change but not systematic loss. *Science* 344:296–299.

Drakare, S., J. J. Lennon, and H. Hillebrand. 2006. The imprint of the geographical, evolutionary and ecological context on species-area relationships. *Ecology Letters* 9:215–227.

Dray, S., and P. Legendre. 2008. Testing the species traits-environment relationships: The fourth-corner problem revisited. *Ecology* 89:3400–3412.

Drummond, A. J., M. A. Suchard, D. Xie, and A. Rambaut. 2012. Bayesian phylogenetics with BEAUti and the BEAST 1.7. *Molecular Biology And Evolution* 19:1969–1973.

Duncan, R. P., and P. A. Williams. 2002. Ecology: Darwin's naturalization hypothesis challenged. *Nature* 417:608–609.

Edgar, R. C. 2004. MUSCLE: Multiple sequence alignment with high accuracy and high throughput. *Nucleic Acids Research* 32:1792–1797.

Edgington, E. S. 1987. *Randomization Tests*. Marcel Dekker, New York, NY.

Edwards, E. J., C. J. Still, and M. J. Donoghue. 2007. The relevance of phylogeny to studies of global change. *Trends in Ecology & Evolution* 22:243–249.

Ehrlich, P. R., and A. H. Ehrlich. 1992. The value of biodiversity, *Ambio* 21:219–226.

Elton, C. S. 1946. Competition and the structure of ecological communities. *Journal of Animal Ecology* 15:54–68.

Elton, C. S. 1958. *The Ecology of Invasions by Animals and Plants*. Maltheun, London.

Etienne, R. S., and J. Rosindell. 2012. Prolonging the past counteracts the pull of the present: Protracted speciation can explain observed slowdowns in diversification. *Systematic Biology* 61:204–213.

Evans, M.E.K., S. A. Smith, R. S. Flynn, and M. J. Donoghue. 2009. Climate, niche evolution, and diversification of the "Bird Cage" evening primroses (Oenothera, sections Anogra and Kleinia). *American Naturalist* 173:225–240.

Faith, D. P. 1992a. Conservation evaluation and phylogenetic diversity. *Biological Conservation* 61:1–10.

Faith, D. P. 1992b. Systematics and conservation: On predicting the feature diversity of subsets of taxa. *Cladistics* 8:361–373.

Faith, D. P. 1994. Phylogenetic pattern and the quantification of organismal biodiversity. *Philosophical Transactions of the Royal Society of London B, Biological Sciences* 345:45–58.

Faith, D. P. 2002. Quantifying biodiversity: A phylogenetic perspective. *Conservation Biology* 16:248–252.

Faith, D. P., S. Magallón, A. P. Hendry, E. Conti, T. Yahara, and M. J. Donoghue. 2010. Evosystem services: An evolutionary perspective on the links between biodiversity and human well-being. *Current Opinion in Environmental Sustainability* 2:66–74.

Faith, D. P., and P. A. Walker. 1996. Integrating conservation and development: Incorporating vulnerability into biodiversity-assessment of areas. *Biodiversity and Conservation* 5:417–429.

Felsenstein, J. 1978. The number of evolutionary trees. *Systematic Biology* 27:27–33.

Felsenstein, J. 1985. Phylogenies and the comparative method. *American Naturalist* 125:1–15.

Felsenstein, J. 2004. *Inferring Phylogenies*. Sinauer Associates, Sunderland, MA, US.

Feng, G., J. Zhang, N. Pei, M. Rao, X. Mi, H. Ren, and K. Ma. 2012. Comparison of phylobetadiversity indices based on community data from Gutianshan forest plot. *Chinese Science Bulletin* 57:623–630.

Ferrier, S., G. Manion, J. Elith, and K. Richardson. 2007. Using generalized dissimilarity modelling to analyse and predict patterns of beta diversity in regional biodiversity assessment. *Diversity and Distributions* 13:252–264.

FitzJohn, R. G. 2010. Quantitative traits and diversification. *Systematic Biology* 59: 619–633.

FitzJohn, R. G. 2012. Diversitree: Comparative phylogenetic analyses of diversification in R. *Methods in Ecology and Evolution* 3:1084–1092.

Fitzpatrick, B. M., and M. Turelli. 2006. The geography of mammalian speciation: Mixed signals from phylogenies and range maps. *Evolution* 60:601–615.

Forest, F., R. Grenyer, M. Rouget, T. J. Davies, R. M. Cowling, D. P. Faith, A. Balmford, et al. 2007. Preserving the evolutionary potential of floras in biodiversity hotspots. *Nature* 445:757–760.

Fox, B. J., and J. H. Brown. 1993. Assembly rules for functional-groups in North-American desert rodent communities. *Oikos* 67:358–370.

Freckleton, R. P. 2000. Phylogenetic tests of ecological and evolutionary hypotheses: Checking for phylogenetic independence. *Functional Ecology* 14:129–134.

Freckleton, R. P. 2009. The seven deadly sins of comparative analysis. *Journal of Evolutionary Biology* 22:1367–1375.

Fritz, S. A., O. R. Bininda-Emonds, and A. Purvis. 2009. Geographical variation in predictors of mammalian extinction risk: Big is bad, but only in the tropics. *Ecology Letters* 12:538–549.

Fritz, S. A., and A. Purvis. 2010. Selectivity in mammalian extinction risk and threat types: A new measure of phylogenetic signal strength in binary traits. *Conservation Biology* 24:1042–1051.

Fukami, T., H.J.E. Beaumont, X.-X. Zhang, and P. B. Rainey. 2007. Immigration history controls diversification in experimental adaptive radiation. *Nature* 446:436–439.

Futuyma, D. J. 2010. Evolutionary constraint and ecological consequences. *Evolution* 64:1865–1884.

Garland, T., P. E. Midford, and A. R. Ives. 1999. An introduction to phylogenetically based statistical methods, with a new method for confidence intervals on ancestral values. *American Zoologist* 39:374–388.

Gause, G. F. 1934. *The Struggle for Existence*. Williams and Wilkins, Baltimore.

Gerhold, P., J. F. Cahill, M. Winter, I. V. Bartish, and A. Prinzing. 2015. Phylogenetic patterns are not proxies of community assembly mechanisms (they are far better). *Functional Ecology* 29:600–614.

Gilpin, M. E., and J. M. Diamond. 1984. Are species co-occurrences on islands non-random, and are null hypotheses useful in community ecology? in *Ecological Communities: Conceptual Issues and the Evidence*, ed. D. R. Strong, D. Simberloff, L. G. Abele, and A. Thistle, pp. 297–315. Princeton University Press, Princeton, NJ, US.

Gittleman, J. L., C. G. Anderson, M. Kot, and H.-K. Luh. 1996. Phylogenetic lability and rates of evolution: A comparison of behavioral, morphological and life history traits, in *Phylogenies and the Comparative Method in Animal Behavior*, ed. E. P. Martins, pp. 166–205. Oxford University Press, Oxford, UK.

Gleason, H. A. 1922. On the relation between species and area. *Ecology* 3:158–162.

Godfray, H.C.J., and J. H. Lawton. 2001. Scale and species numbers. *Trends in Ecology & Evolution* 16:400–404.

Godoy, O., N.J.B. Kraft, and J. M. Levine. 2014. Phylogenetic relatedness and the determinants of competitive outcomes. *Ecology Letters* 17:836–844.

Gotelli, N. J. 2000. Null model analysis of species co-occurrence patterns. *Ecology* 81:2606–2621.

Gotelli, N. J. 2001. Research frontiers in null model analysis. *Global Ecology and Biogeography* 10:337–343.

Gotelli, N. J., and G. R. Graves. 1996. *Null Models in Ecology*. Smithsonian Institution Press, Washington, DC.

Gould, S. J., and N. Eldredge. 1977. Punctuated equilibria: The tempo and mode of evolution reconsidered. *Paleobiology* 3:115–151.

Graham, C. H., and P.V.A. Fine. 2008. Phylogenetic beta diversity: Linking ecological and evolutionary processes across space in time. *Ecology Letters* 11:1265–1277.

Graham, C. H., J. L. Parra, C. Rahbek, and J. A. McGuire. 2009. Phylogenetic structure in tropical hummingbird communities. *Proceedings of the National Academy of Sciences of the United States of America* 106:19673–19678.

Grant, P. R. 1966. Ecological compatibility of bird species on islands. *American Naturalist* 100:451–462.

Grant, P. R. 1972. Convergent and divergent character displacement. *Biological Journal of the Linnean Society* 4:39–68.

Gravel, D., C. D. Canham, M. Beaudet, and C. Messier. 2006. Reconciling niche and neutrality: The continuum hypothesis. *Ecology Letters* 9:399–409.

Graves, J.A.M., and M. B. Renfree. 2013. Marsupials in the age of genomics. *Annual Review of Genomics and Human Genetics* 14:393–420.

Grenyer, R., C.D.L. Orme, S. F. Jackson, G. H. Thomas, R. G. Davies, T. J. Davies, K. E. Jones, et al. 2006. Global distribution and conservation of rare and threatened vertebrates. *Nature* 444:93–96.

Grime, J. P. 1979. *Plant Strategies and Vegetation Processes*. Wiley, Chichester, UK.

Grinnell, J. 1917. The niche relationships of the California thrasher. *The Auk* 34:427–433.

Grinnell, J. 1925. Risks incurred in the introduction of alien game birds. *Science* 61:621–623.

Gruter, D., B. Schmid, and H. Brandl. 2006. Influence of plant diversity and elevated atmospheric carbon dioxide levels on belowground bacterial diversity. *BMC Microbiology* 27:68–76.

Guillot, G., and F. Rousset. 2013. Dismantling the Mantel tests. *Methods in Ecology and Evolution* 4:336–344.

Guindon, S., J. F. Dufayard, V. Lefort, M. Anisimova, W. Hordijk, and O. Gascuel. 2010. New algorithms and methods to estimate maximum-likelihood phylogenies: Assessing the performance of PhyML 3.0. *Systematic Biology* 59:307–321.

Haberl, H., K. H. Erb, F. Krausmann, V. Gaube, A. Bondeau, C. Plutzar, S. Gingrich, et al. 2007. Quantifying and mapping the human appropriation of net primary production in earth's terrestrial ecosystems. *Proceedings of the National Academy of Sciences of the United States of America* 104:12942–12947.

Halpern, B. S., C. R. Pyke, H. E. Fox, J. Chris Haney, M. A. Schlaepfer, and P. Zaradic. 2006. Gaps and mismatches between global conservation priorities and spending. *Conservation Biology* 20:56–64.

Hansen, T. F. 1997. Stabilizing selection and the comparative analysis of adaptation. *Evolution* 51:1341–1351.

Hardy, O. J. 2008. Testing the spatial phylogenetic structure of local communities: Statistical performances of different null models and test statistics on a locally neutral community. *Journal of Ecology* 96:914–926.

Hardy, O. J., and S. Pavoine. 2012. Assessing phylogenetic signal with measurement error: A comparison of Mantel tests, Blomberg et al.'s K, and phylogenetic distograms. *Evolution* 66:2614–2621.

Hardy, O. J., and B. Senterre. 2007. Characterizing the phylogenetic structure of communities by an additive partitioning of phylogenetic diversity. *Journal of Ecology* 95:493–506.

Harmon, L. J., J. B. Losos, T. J. Davies, R. G. Gillespie, J. L. Gittleman, W. B. Jennings, K. H. Kozak, et al. 2010. Early bursts of body size and shape evolution are rare in comparative data. *Evolution* 64:2385–2396.

Harmon, L. J., J. T. Weir, C. D. Brock, R. E. Glor, and E. Challanger. 2008. GEIGER: Investigating evolutionary radiations. *Bioinformatics* 24:129–131.

Harmon-Threatt, A. N., and D. D. Ackerly. 2013. Filtering across spatial scales: Phylogeny, biogeography and community structure in bumble bees. *PLoS ONE* 8:e60446.

Harper, J. L. 1978. *Population Biology of Plants*. Academic Press, London.

Harvey, P. H., R. K. Colwell, J. W. Silvertown, and R. M. May. 1983. Null models in ecology. *Annual Review of Ecology and Systematics* 14:189–211.

Harvey, P. H., and M. D. Pagel. 1991. *The Comparative Method in Evolutionary Biology*. Oxford University Press, Oxford, UK.

Harvey, P. H., A. F. Read, and S. Nee. 1995a. Further remarks on the role of phylogeny in comparative ecology. *Journal of Ecology* 83:733–734.

Harvey, P. H., A. F. Read, and S. Nee. 1995b. Why ecologists need to be phylogenetically challenged. *Journal of Ecology* 83:535–536.

Heard, S. B. 1992. Patterns in tree balance among cladistic, phenetic, and randomly generated phylogenetic trees. *Evolution* 46:1818–1826.

Heard, S. B. 1996. Patterns in phylogenetic tree balance with variable and evolving speciation rates. *Evolution* 50:2141–2148.

Heard, S. B., and A. Ø. Mooers. 2000. Phylogenetically patterned speciation rates and extinction risks change the loss of evolutionary history during extinctions. *Proceedings of the Royal Society of London B, Biological Sciences* 267:613–620.

Heard, S. B., and A. Ø. Mooers. 2002. Signatures of random and selective mass extinctions in phylogenetic tree balance. *Systematic Biology* 51:889–897.

Hector, A., K. Dobson, A. Minns, E. Bazeley-White, and J. Hartley Lawton. 2001. Community diversity and invasion resistance: An experimental test in a grassland ecosystem and a review of comparable studies. *Ecological Research* 16:819–831.

Heip, C.H.R., P.M.J. Herman, and K. Soetart. 1998. Indices of diversity and evenness. *Oceanis* 24:61–87.

Helmus, M. R., T. J. Bland, C. K. Williams, and A. R. Ives. 2007. Phylogenetic measures of biodiversity. *American Naturalist* 169:E68–E83.

Helmus, M. R., and A. R. Ives. 2012. Phylogenetic diversity–area curves. *Ecology* 93:S31–S43.

Helmus, M. R., W. Keller, M. J. Paterson, N. D. Yan, C. H. Cannon, and J. A. Rusak. 2010. Communities contain closely related species during ecosystem disturbance. *Ecology Letters* 13:162–174.

Helmus, M. R., K. Savage, M. W. Diebel, J. T. Maxted, and A. R. Ives. 2007. Separating the determinants of phylogenetic community structure. *Ecology Letters* 10:917–925.

Hill, M. O. 1973. Diversity and evenness: A unifying notation and its consequences. *Ecology* 54:427–432.

Hillebrand, H. 2004. On the generality of the latitudinal diversity gradient. *American Naturalist* 163:192–211.

Holder, M., and P. O. Lewis. 2003. Phylogeny estimation: Traditional and Bayesian approaches. *Nature Reviews Genetics* 4:275–284.

Holling, C. S. 1992. Cross-scale morphology, geometry, and dynamics of ecosystems. *Ecological Monographs* 62:447–502.

Holmes, R. T., and F. A. Pitelka. 1968. Food overlap among coexisting sandpipers on northern Alaskan tundra. *Systematic Zoology* 17:305–318.

Holt, R. D. 1996. Demographic constraints in evolution: Towards unifying the evolutionary theories of senescence and niche conservatism. *Evolutionary Ecology* 10:1–11.

Holt, R. D. 2006. Emergent neutrality. *Trends in Ecology & Evolution* 21:531–533.

Holt, R. D., J. H. Lawton, G. A. Polis, and N. D. Martinez. 1999. Trophic rank and the species-area relationship. *Ecology* 80:1495–1504.

Holyoak, M., M. A. Leibold, and R. D. Holt, eds. 2005. *Metacommunities: Spatial Dynamics and Ecological Communities*. University of Chicago Press, Chicago, IL.

Huang, S., T. J. Davies, and J. L. Gittleman. 2011. How global extinctions impact regional biodiversity in mammals. *Biology Letters* 8:222-225.

Hubbell, S. P. 2001. *The Unified Neutral Theory of Biodiveristy and Biogeography*. Princeton University Press, Princeton, NJ, US.

Huelsenbeck, J. P., R. Nielsen, and J. P. Bollback. 2003. Stochastic mapping of morphological characters. *Systematic Biology* 52:131–158.

Huelsenbeck, J. P., and F. R. Ronquist. 2001. MRBAYES: Bayesian inference of phylogenetic trees. *Bioinformatics* 17:754–755.

Hurlbert, S. H. 1984. Pseudoreplication and the design of ecological field experiments. *Ecological Monographs* 54:187–211.

Hutchinson, G. E. 1959. Homage to Santa Rosalia or why are there so many kinds of species? *American Naturalist* 93:145–159.

Ings, T. C., J. M. Montoya, J. Bascompte, N. Bluthgen, L. Brown, C. F. Dormann, F. Edwards, et al. 2009. Ecological networks: Beyond food webs. *Journal of Animal Ecology* 78:253–269.

Isaac, N.J.B., J. Mallet, and G. M. Mace. 2004. Taxonomic inflation: Its influence on macroecology and conservation. *Trends in Ecology & Evolution* 19:464–469.

Isaac, N.J.B., S. T. Turvey, B. Collen, C. Waterman, and J.E.M. Baillie. 2007. Mammals on the EDGE: Conservation priorities based on threat and phylogeny. *PLoS ONE* 2:e296.

Izsák, J., and L. Papp. 2000. A link between ecological diversity indices and measures of biodiversity. *Ecological Modelling* 130:151–156.

Jaccard, P. 1901. Étude comparative de la distribution florale dans une portion des Alpes et du Jura. *Bulletin de la Société Vaudoise des Sciences Naturelle* 37:547–579.

Jaccard, P. 1922. La chorologie sélective et sa signification pour la sociologie végétale. *Mémoires de la Société Vaudoise des Sciences Naturelle* 2:81–107.

Jackson, D. 1995. PROTEST: A PROcrustean randomization TEST of community environment concordance. *Ecoscience* 2:297–303.

James, A., K. J. Gaston, and A. Balmford. 2001. Can we afford to conserve biodiversity? *Bioscience* 51:43–52.

Jarvinen, O. 1982. Species-to-genus ratios in biogeography: A historical note. *Journal of Biogeography* 9:363–370.

Jetz, W., G. Thomas, J. Joy, K. Hartmann, and A. Mooers. 2012. The global diversity of birds in space and time. *Nature* 491:444–448.

Jiang, L., J. Q. Tan, and Z. C. Pu. 2010. An experimental test of Darwin's naturalization hypothesis. *American Naturalist* 175:415–423.

Jin, L. S., M. W. Cadotte, and M.-J. Fortin. 2015. Phylogenetic turnover implicates niche conservatism in montane plant species. *Journal of Ecology* 103:742-749.

Johnson, M.T.J., and J. R. Stinchcombe. 2007. An emerging synthesis between community ecology and evolutionary biology. *Trends in Ecology & Evolution* 22:250–257.

Jost, L., A. Chao, and R. L. Chazdon. 2011. Compositional similarity and beta diversity, in *Biological Diversity*, ed. A. E. Magurran and B. J. McGill, pp. 66–84. Oxford University Press, Oxford, UK.

Kareiva, P., and S. A. Levin, eds. 2003. *The Importance of Species*. Princeton University Press, Princeton, NJ.

Kareiva, P., and M. Marvier. 2003. Conserving biodiversity coldspots. *American Scientist* 91:344–351.

Katoh, K., K. Misawa, K. I. Kuma, and T. Miyata. 2002. MAFFT: A novel method for rapid multiple sequence alignment based on fast Fourier transform. *Nucleic Acids Research* 30:3059–3066.

Keddy, P. A. 1992. Assembly and response rules: Two goals for predictive community ecology. *Journal of Vegetation Science* 3:157–164.

Kelly, S., R. Grenyer, and R. W. Scotland. 2014. Phylogenetic trees do not reliably predict feature diversity. *Diversity and Distributions* 20:600–612.

Kembel, S. W. 2009. Disentangling niche and neutral influences on community assembly: Assessing the performance of community phylogenetic structure tests. *Ecology Letters* 12:949–960.

Kembel, S. W., P. D. Cowan, M. R. Helmus, W. K. Cornwell, H. Morlon, D. D. Ackerly, S. P. Blomberg, and C. O. Webb. 2010. Picante: R tools for integrating phylogenies and ecology. *Bioinformatics* 26:1463–1464.

King, W. 1685. Of the bogs, and loughs of Ireland by Mr. William King, Fellow of the Dublin Society, as it was presented to that society. *Philosophical Transactions of the Royal Society of London* 15:948–960.

Kirkpatrick, J. B. 1983. An iterative method for establishing priorities for the selection of nature reserves: An example from Tasmania. *Biological Conservation* 25:127–134.

Koh, L. P., R. R. Dunn, N. S. Sodhi, R. K. Colwell, H. C. Proctor, and V. S. Smith. 2004. Species coextinctions and the biodiversity crisis. *Science* 305:1632–1634.

Kohler, R. E. 2002. *Landscapes and Labscapes: Exploring the Lab-Field Border in Biology.* University of Chicago Press, Chicago, IL, US.

Konecny, M. J. 1989. Movement patterns and food habits of four sympatric carnivore species in Belize, Central America. *Advances in Neotropical Mammalogy* 1989:243–264.

Kozak, K., and J. J. Wiens. 2010. Niche conservatism drives elevational diversity patterns in Appalachian salamanders. *American Naturalist* 176:40–54.

Kraft, N., and D. Ackerly. 2010. Functional trait and phylogenetic tests of community assembly across spatial scales in an Amazonian forest. *Ecological Monographs* 80:401–422.

Kraft, N.J.B., W. K. Cornwell, C. O. Webb, and D. D. Ackerly. 2007. Trait evolution, community assembly and the phylogenetic structure of ecological communities. *American Naturalist* 170:271–283.

Kraft, N.J.B., O. Godoy, and J. M. Levine. 2015. Plant functional traits and the multidimensional nature of species coexistence. *Proceedings of the National Academy of Science of the United States of America* 112:797–802.

Kuhn, T. S., A. Ø. Mooers, and G. H. Thomas. 2011. A simple polytomy resolver for dated phylogenies. *Methods in Ecology and Evolution* 2:427–436.

Lambdon, P. W., and P. E. Hulme. 2006. Predicting the invasion success of Mediterranean alien plants from their introduction characteristics. *Ecography* 29:853–865.

Lande, R. 1996. Statistics and partitioning of species diversity, and similarity among multiple communities. *Oikos* 76:5–13.

Lanier, H. C., D. L. Edwards, and L. L. Knowles. 2013. Phylogenetic structure of vertebrate communities across the Australian arid zone. *Journal of Biogeography* 40:1059–1070.

Larkin, M. A., G. Blackshields, N. Brown, R. Chenna, P. A. McGettigan, H. McWilliam, F. Valentin et al. 2007. Clustal W and Clustal X version 2.0. *Bioinformatics* 23:2947–2948.

Lavergne, S., M. E. Evans, I. J. Burfield, F. Jiguet, and W. Thuiller. 2013. Are species' responses to global change predicted by past niche evolution? *Philosophical Transactions of the Royal Society of London B, Biological Sciences* 368:20120091.

Lawing, A. M., and P. D. Polly. 2011. Pleistocene climate, phylogeny, and climate envelope models: An integrative approach to better understand species' response to climate change. *PLoS ONE* 6:e28554.

Legendre, P., Y. Desdevises, and E. Bazin. 2002. A statistical test for host–parasite coevolution. *Systematic Biology* 51:217–234.

Legendre, P., and M.-J. Fortin. 2010. Comparison of the Mantel test and alternative approaches for detecting complex multivariate relationships in the spatial analysis of genetic data. *Molecular Ecology Resources* 10:831–844.

Leibold, M. A., M. Holyoak, N. Mouquet, P. Amarasekare, J. M. Chase, M. F. Hoopes, R. D. Holt, et al. 2004. The metacommunity concept: A framework for multi-scale community ecology. *Ecology Letters* 7:601–613.

Leinster, T., and C. A. Cobbold. 2011. Measuring diversity: The importance of species similarity. *Ecology* 93:477–489.

Lessard, J.-P., J. Belmaker, J. A. Myers, J. M. Chase, and C. Rahbek. 2012. Inferring local ecological processes amid species pool influences. *Trends in Ecology & Evolution* 27:600–607.

Lessard, J.-P., M. K. Borregaard, J. A. Fordyce, C. Rahbek, M. D. Weiser, R. R. Dunn, and N. J. Sanders. 2012. Strong influence of regional species pools on continent-wide structuring of local communities. *Proceedings of the Royal Society of London B, Biological Sciences* 279:266–274.

Lessard-Therrien, M., T. J. Davies, and K. Bolmgren. 2014. A phylogenetic comparative study of flowering phenology along an elevational gradient in the Canadian subarctic. *International Journal of Biometeorology* 58:455–462.

Letten, A. D., and W. K. Cornwell. 2014. Trees, branches and (square) roots: Why evolutionary relatedness is not linearly related to functional distance. *Methods in Ecology and Evolution* 6:439–444.

Levine, J. M. 2000. Species diversity and biological invasions: Relating local process to community pattern. *Science* 288:852–854.

Lewin, R. 1983. Santa Rosalia was a goat: Ecologists have for two decades made assumptions about the importance of competition in community organization; that idea is now under vigorous attack. *Science* 221:636–639.

Lipman, D. J., and W. R. Pearson. 1985. Rapid and sensitive protein similarity searches. *Science* 227:1435–1441.

Lippmaa, T. 1939. The unistratal concept of plant communities (the unions). *American Midland Naturalist* 21:111–145.

Liu, X., M. Liang, R. S. Etienne, Y. Wang, C. Staehelim, and S. Yu. 2011. Experimental evidence for a phylogenetic Janzen-Connell effect in a subtropical forest. *Ecology Letters* 15:111–118.

Liu, J., X. Zhang, F. Song, S. Zhou, M. Cadotte, and C.J.A. Bradshaw. 2015. Explaining maximum variation in productivity requires phylogenetic diversity and single functional traits. *Ecology* 96:176–183.

Lortie, C. J., R. W. Brooker, P. Choler, Z. Kikvidze, R. Michalet, F. I. Pugnaire, and R. M. Callaway. 2004. Rethinking plant community theory. *Oikos* 107:433–438.

Losos, J. B., and R. E. Glor. 2003. Phylogenetic comparative methods and the geography of speciation. *Trends in Ecology & Evolution* 18:220–227.

Loveridge, A. J., and D. W. Macdonald. 2002. Habitat ecology of two sympatric species of jackals in Zimbabwe. *Journal of Mammalogy* 83:599–607.

Loveridge, A. J., and D. W. Macdonald. 2003. Niche separation in sympatric jackals (*Canis nesomelas* and *Canis adustus*). *Journal of Zoology* 259:143–153.

Lovette, I. J., and W. M. Hochachka. 2006. Simultaneous effects of phylogenetic niche conservatism and competition on avian community structure. *Ecology* 87:S14–S28.

Lozupone, C., and R. Knight. 2005. UniFrac: A new phylogenetic method for comparing microbial communities. *Applied and Environmental Microbiology* 71:8228–8235.

Lunneborg, C. E. 2000. Random assignment of available cases: Let the inferences fit the design. Unpublished work, University of Washington, Seattle, WA.

Lynch, J. D. 1989. The gauge of speciation: On the frequency of modes of speciation, in *Speciation and Its Consequences*, ed. D. Otte and J. A. Endler, pp. 527–553. Sinauer, Sunderland, MA, US.

MacArthur, R., and R. Levins. 1967. The limiting similarity, convergence, and divergence of coexisting species. *American Naturalist* 101:377–385.

MacArthur, R. H. 1958. Population ecology of some warblers of Northeastern coniferous forests. *Ecology* 39:599–619.

MacArthur, R. H., and E. O. Wilson. 1967. *The Theory of Island Biogeography*. Princeton University Press, Princeton, NJ, US.

Mace, G. M., A. Balmford, L. Boitani, G. Cowlishaw, A. Dobson, D. Faith, K. J. Gaston, et al. 2000. It's time to work together and stop duplicating conservation efforts . . . Nature 405:393–393.

Mace, G. M., J. L. Gittleman, and A. Purvis. 2003. Preserving the tree of life. *Science* 300:1707–1709.

Mace, G. M., H. Masundire, and J. Baillie. 2005. Biodiversity, in *Ecosystems and Human Well-Being: Current State and Trends*, ed. R. Hassan, R. J. Scholes, and N. Ash, pp. 77–122. Island Press, Washington, DC.

Maddison, W. P. 2006. Confounding asymmetries in evolutionary diversification and character change. *Evolution* 60:1743–1746.

Maddison, W. P., and D. R. Maddison. 2015. Mesquite: A modular system for evolutionary analysis. Version 3.04 http://mesquiteproject.org.

Maddison, W. P., P. E. Midford, and S. P. Otto. 2007. Estimating a binary character's effect on speciation and extinction. *Systematic Biology* 56:701–710.

Maestre, F., R. M. Callaway, F. Valladares, and C. J. Lortie. 2009. Refining the stress-gradient hypothesis for competition and facilitation in plant communities. *Journal of Ecology* 97:199–205.

Magurran, A. E. 2003. *Measuring Biological Diversity.* Blackwell Science, Oxford, UK.

Maherali, H., and J. N. Klironomos. 2007. Influence of phylogeny on fungal community assembly and ecosystem functioning. *Science* 316:1746–1748.

Maillefer, A. 1928. Les courbes de Willis: Répartition des espèces dans les genres de différente étendue. *Bulletin de la Société Vaudoise des Sciences Naturelle* 56:617–631.

Maillefer, A. 1929. Le Coefficient générique de P. Jaccard et sa signification. *Mémoires de la Société Vaudoise des Sciences Naturelle* 3:9–183.

Manly, B.F.J. 1991. *Randomization, Bootstrap and Monte Carlo Methods in Biology.* Chapman and Hall, London.

Mannion, P. D., P. Upchurch, R. B. Benson, and A. Goswami. 2014. The latitudinal biodiversity gradient through deep time. *Trends in Ecology & Evolution* 29:42–50.

Margules, C. R., A. O. Nicholls, and R. L. Pressey. 1988. Selecting networks of reserves to maximise biological diversity. *Biological Conservation* 43:63–76.

Margules, C. R., and R. L. Pressey. 2000. Systematic conservation planning. *Science* 405:243–253.

Martinez-Meyer, E., and A. T. Peterson. 2006. Conservatism of ecological niche characteristics in North American plant species over the Pleistocene-to-Recent transition. *Journal of Biogeography* 33:1779–1789.

Martins, E. P., and T. F. Hansen. 1997. Phylogenies and the comparative method: A general approach to incorporating phylogenetic information into the analysis of interspecific data. *American Naturalist* 149:646–667.

Mattern, M. Y., and D. A. McLennan. 2000. Phylogeny and speciation of felids. *Cladistics* 16:232–253.

May, R. M. 1981. *Theoretical Ecology: Principles and Applications.* Sinauer, Oxford, UK.

Mayfield, M. M., and J. M. Levine. 2010. Opposing effects of competitive exclusion on the phylogenetic structure of communities. *Ecology Letters* 13:1085–1093.

Mayr, E. 1942. *Systematics and the Origin of Species from the Viewpoint of a Zoologist.* Harvard University Press, Cambridge, MA, US.

Mayr, E. 1963. *Animal Speciation and Evolution.* Harvard University Press, Cambridge, MA, US.

Mazel, F., F. Guilhaumon, N. Mouquet, V. Devictor, D. Gravel, J. Renaud, M. V. Cianciaruso, et al. 2014. Multifaceted diversity-area relationships reveal global hotspots of mammalian species, trait and lineage diversity. *Global Ecology and Biogeography* 23:836–847.

Mazer, S. J., S. E. Travers, B. I. Cook, T. J. Davies, K. Bolmgren, N. J. Kraft, N. Salamin, and D. W. Inouye. 2013. Flowering date of taxonomic families predicts phenological sensitivity to temperature: Implications for forecasting the effects of climate change on unstudied taxa. *American Journal of Botany* 100:1381–1397.

McGill, B. J., B. J. Enquist, E. Weiher, and M. Westoby. 2006. Rebuilding community ecology from functional traits. *Trends in Ecology & Evolution* 21:178–185.

McKinney, M. L. 2004. Measuring floristic homogenization by non-native plants in North America. *Global Ecology and Biogeography* 13:47–53.

McKinney, M. L. 2006. Urbanization as a major cause of biotic homogenization. *Biological Conservation* 127:247–260.

McKinney, M. L., and F. A. La Sorte. 2007. Invasiveness and homogenization: Synergism of wide dispersal and high local abundance. *Global Ecology and Biogeography* 16:394–400.

McKinney, M. L., and J. L. Lockwood. 1999. Biotic homogenization: A few winners replacing many losers in the next mass extinction. *Trends in Ecology & Evolution* 14:450–453.

Miklós, I., and J. Podani. 2004. Randomization of presence–absence matrices: Comments and new algorithms. *Ecology* 85:86–92.

Mittelbach, G. G., D. W. Schemske, H. V. Cornell, A. P. Allen, J. M. Brown, M. B. Bush, S. P. Harrison, et al. 2007. Evolution and the latitudinal diversity gradient: Speciation, extinction and biogeography. *Ecology Letters* 10:315–331.

Mittermeier, R. A., G. P. Robles, M. Hoffmann, J. D. Pilgrim, T. M. Brooks, C. G. Mittermeier, J. F. Lamoreux, and G.A.B. da Fonseca. 2004. *Hotspots Revisited*. CEMEX, Mexico City, Mexico.

Miyamoto, M. M., and J. Cracraft. 1991. *Phylogenetic Analysis of DNA Sequences*. Oxford University Press, Oxford, UK.

Mokany, K., J. Ash, and S. Roxburgh. 2008. Functional identity is more important than diversity in influencing ecosystem processes in a temperate native grassland. *Journal of Ecology* 96:884–893.

Mooers, A. Ø., L. J. Harmon, M.G.B. Blum, D.H.J. Wong, and S. B. Heard. 2007. Some models of phylogenetic tree shape, in *Reconstructing Evolution: New Mathematical and Computational Advances*, ed. O. Gascuel and M. Steel, pp. 149–170. Oxford University Press, Oxford, NY.

Mooers, A. Ø., S. B. Heard, and E. Chrostowski. 2005. Evolutionary heritage as a metric for conservation, in *Phylogeny and Conservation*, ed. A. Purvis, T. L. Brooks, and J. L. Gittleman, pp. 120–138. Oxford University Press, Oxford, UK.

Moreau, R. E. 1966. *The Bird Faunas of Africa and Its Islands*. Academic Press, New York, NY.

Morlon, H., D. W. Schwilk, J. A. Bryant, P. A. Marquet, A. G. Rebelo, C. Tauss, B.J.M. Bohannan, and J. L. Green. 2011. Spatial patterns of phylogenetic diversity. *Ecology Letters* 14:141–149.

Mouchet, M. A., and D. Mouillot. 2011. Decomposing phylogenetic entropy into alpha, beta and gamma components. *Biology Letters* 7:205–209.

Mouquet, N., V. Devictor, C. Meynard, F. Munoz, L. F. Bersier, J. Chave, P. Couteron, et al. 2012. Ecophylogenetics: Advances and perspectives. *Biological Reviews* 87:769–785.

Muenkemueller, T., L. Gallien, S. Lavergne, J. Renaud, C. Roquet, S. Abdulhak, S. Dullinger, et al. 2014. Scale decisions can reverse conclusions on community assembly processes. *Global Ecology and Biogeography* 23:620–632.

Myers, N., R. A. Mittermeier, C. G. Mittermeier, G.A.B. da Fonseca, and J. Kent. 2000. Biodiversity hotspots for conservation priorities. *Nature* 403:853–858.

Myster, R. W., and S.T.A. Pickett. 1990. Initial conditions, history and successional pathways in 10 contrasting old fields. *American Midland Naturalist* 124:231–238.

Naeem, S. 2002. Disentangling the impacts of diversity on ecosystem functioning in combinatorial experiments. *Ecology* 83:2925–2935.

Narwani, A., M. A. Alexandrou, T. H. Oakley, I. T. Carroll, and B. J. Cardinale. 2013. Experimental evidence that evolutionary relatedness does not affect the ecological mechanisms of coexistence in freshwater green algae. *Ecology Letters* 16:1373–1381.

Nee, S., and R. M. May. 1997. Extinction and the loss of evolutionary history. *Science* 278:692–694.

Nee, S., R. M. May, and P. H. Harvey. 1994. The reconstructed evolutionary process. *Philosophical Transactions of the Royal Society of London B, Biological Sciences* 344:305–311.

Nielsen, R. 2002. Mapping mutations on phylogenies. *Systematic Biology* 51:729–739.

Nixon, K. C. 1999. The parsimony ratchet, a new method for rapid parsimony analysis. *Cladistics* 15:407–414.

Norf, H., H. Arndt, and M. Weitere. 2007. Impact of local temperature increase on the early development of biofilm-associated ciliate communities. *Oecologia* 151:341–350.

Nowak, R. M. 1999. *Walker's Mammals of the World*. Johns Hopkins University Press, Baltimore, MD.

Nuismer, S. L., and L. J. Harmon. 2015. Predicting rates of interspecific interaction from phylogenetic trees. *Ecology Letters* 18:17–27.

O'Meara, B. C. 2012. Evolutionary inferences from phylogenies: A review of methods. *Annual Review of Ecology, Evolution, and Systematics* 43:267–285.

O'Meara, B. C., C. Ané, M. J. Sanderson, P. C. Wainwright, and T. Hansen. 2006. Testing for different rates of continuous trait evolution using likelihood. *Evolution* 60:922–933.

Oke, O. A., S. B. Heard, and J. T. Lundholm. 2014. Integrating phylogenetic community structure with species distribution models: An example with plants of rock barrens. *Ecography* 37:614–625.

Oksanen, J., R. Kindt, P. Legendre, R. O'Hara, G. L. Simpson, M.H.H. Stevens, and H. Wagner. 2008. Vegan: Community ecology package. http://cran.r-project.org, https://github.com/vegandevs/vegan.

Olden, J. D., and T. P. Rooney. 2006. On defining and quantifying biotic homogenization. *Global Ecology and Biogeography* 15:113–120.

Owens, I. P., and P. M. Bennett. 2000. Ecological basis of extinction risk in birds: Habitat loss versus human persecution and introduced predators. *Proceedings of the National Academy of Sciences of the United States of America* 97:12144–12148.

Pacala, R. W., and M. Rees. 1998. Models suggesting field experiments to test two hypotheses explaining successional diversity. *American Naturalist* 152:729–737.

Pagel, M. 1994. Detecting correlated evolution on phylogenies: A general method for the comparative analysis of discrete characters. *Proceedings of the Royal Society of London B, Biological Sciences* 255:37–45.

Pagel, M. 1997. Inferring evolutionary processes from phylogenies. *Zoologica Scripta* 26:331–348.

Pagel, M. 1999. Inferring the historical patterns of biological evolution. *Nature* 401:877–884.

Palmgren, A. 1921. Die Entfernung als pflanzengeographischer faktor. *Series Acta Societatis pro Fauna et Flora Fennica* 49:1–113.

Palmgren, A. 1925. Die Artenzahl als pflanzengeographischer Charakter sowie der Zufall und die säkulare Landhebung als pflanzengeographischer Faktoren. Ein pflanzengeographische Entwurf, basiert auf Material aus dem åländischen Schärenarchipel. *Acta Botanica Fennica* 1:1–143.

Palomares, F., and T. M. Caro. 1999. Interspecific killing among mammalian carnivores. *American Naturalist* 153:492–508.

Paradis, E. 2011. *Analysis of Phylogenetics and Evolution with R*. Springer, New York, NY.

Paradis, E., J. Claude, and K. Strimmer. 2004. APE: Analyses of phylogenetics and evolution in R language. *Bioinformatics* 20:289–290.

Parhar, R. K., and A. Ø. Mooers. 2011. Phylogenetically clustered extinction risks do not substantially prune the tree of life. *PLoS ONE* 6:e23528.

Park, D. S., and D. Potter. 2013. A test of Darwin's naturalization hypothesis in the thistle tribe shows that close relatives make bad neighbors. *Proceedings of the National Academy of Sciences of the United States of America* 110:17915–17920.

Parmentier, I., M. Rejou-Mechain, J. Chave, J. Vleminckx, D. W. Thomas, D. Kenfack, G. B. Chuyong, and O. J. Hardy. 2014. Prevalence of phylogenetic clustering at multiple scales in an African rain forest tree community. *Journal of Ecology* 102:1008–1016.

Parmesan, C. 2006. Ecological and evolutionary responses to recent climate change. *Annual Review of Ecology Evolution and Systematics* 37:637–669.

Pavoine, S., and M. Bonsall. 2011. Measuring biodiversity to explain community assembly: A unified approach. *Biological Reviews* 86:792–812.

Pavoine, S., S. Ollier, and A.-B. Dufour. 2005. Is the originality of a species measurable? *Ecology Letters* 8:579–586.

Pavoine, S., E. Vela, S. Gachet, G. De Bélair, and M. B. Bonsall. 2011. Linking patterns in phylogeny, traits, abiotic variables and space: A novel approach to linking environmental filtering and plant community assembly. *Journal of Ecology* 99:165–175.

Pearse, I. S., and A. L. Hipp. 2014. Native plant diversity increases herbivory to non-natives. *Proceedings of the Royal Society of London B, Biological Sciences* 281:20141841.

Pearse, W. D., M. W. Cadotte, J. Cavender-Bares, A. R. Ives, C. M. Tucker, S. C. Walker, and M. R. Helmus. 2015. pez: Phylogenetic for the environmental sciences. *Bioinformatics* (btv277).

Pearse, W. D., J. Cavender-Bares, A. Purvis, and M. R. Helmus. 2014. Metrics and models of community phylogenetics, in *Modern Phylogenetic Comparative Methods and Their Application in Evolutionary Biology*, ed. L. Z. Garamszegi, pp. 451–464. Springer-Verlag, Berlin, Germany.

Pearse, W. D., F. A. Jones, and A. Purvis. 2013. Barro Colorado Island's phylogenetic assemblage structure across fine spatial scales and among clades of different ages. *Ecology* 94:2861–2872.

Pearse, W. D., and A. Purvis. 2013. phyloGenerator: An automated phylogeny generation tool for ecologists. *Methods in Ecology and Evolution* 4:692–698.

Pearson, W. R., and D. J. Lipman. 1988. Improved tools for biological sequence comparison. *Proceedings of the National Academy of Sciences of the United States of America* 85:2444–2448.

Pellissier, L., J.-N. Pradervand, P. H. Williams, G. Litsios, D. Cherix, and A. Guisan. 2013. Phylogenetic relatedness and proboscis length contribute to structuring bumblebee communities in the extremes of abiotic and biotic gradients. *Global Ecology and Biogeography* 22:577–585.

Peres-Neto, P. R., and D. A. Jackson. 2001. How well do multivariate data sets match? The advantages of a Procrustean superimposition approach over the Mantel test. *Oecologia* 129:169–178.

Peres-Neto, P. R., M. A. Leibold, and S. Dray. 2012. Assessing the effects of spatial contingency and environmental filtering on metacommunity phylogenetics. *Ecology* 93:S14–S30.

Peres-Neto, P. R., J. D. Olden, and D. A. Jackson. 2001. Environmentally constrained null models: Site suitability as occupancy criterion. *Oikos* 93:110–120.

Petchey, O. L., and K. J. Gaston. 2002. Functional diversity (FD), species richness and community composition. *Ecology Letters* 5:402–411.

Peterson, D. L., and V. T. Parker. 1998. *Ecological Scale: Theory and Applications.* Columbia University Press, New York, NY.

Phillimore, A. B., and T. D. Price. 2008. Density dependent cladogenesis in birds. *PLoS Biology* 6:e71.

Pillar, V.D.P., and L. Orloci. 1996. On randomization testing in vegetation science: Multifactor comparisons of relevé groups. *Journal of Vegetation Science* 7:585–592.

Pimm, S. L., G. J. Russell, J. L. Gittleman, and T. M. Brooks. 1995. The future of biodiversity. *Science* 269:347–350.

Poffenroth, M., and J. O. Matson. 2007. Habitat partitioning by two sympatric species of chipmunk (genus: *Neotamias*) in the Warner Mountains of California. *Bulletin of the Southern California Academy of Sciences* 106:208–214.

Polis, G. A., C. A. Myers, and R. D. Holt. 1989. The ecology and evolution of intraguild predation. *Annual Review of Ecology and Systematics* 20:297–330.

Posada, D., and K. A. Crandall. 1998. MODELTEST: Testing the model of DNA substitution. *Bioinformatics* 14:817–818.

Pressey, R. L., C. J. Humphries, C. R. Margules, R. I. Vane-Wright, and P. H. Williams. 1993. Beyond opportunism: Key principles for systematic reserve selection. *Trends in Ecology & Evolution* 8:124–128.

Pressey, R. L., H. P. Possingham, and J. R. Day. 1997. Effectiveness of alternative heuristic algorithms for identifying minimum requirements for conservation reserves. *Biological Conservation* 80:207–219.

Proches, S., J.R.U. Wilson, D. M. Richardson, and M. Rejmanek. 2008. Searching for phylogenetic pattern in biological invasions. *Global Ecology and Biogeography* 17:5–10.

Purvis, A., P. M. Agapow, J. L. Gittleman, and G. M. Mace. 2000. Nonrandom extinction and the loss of evolutionary history. *Science* 288:328–330.

Purvis, A., J. L. Gittleman, and T. Brooks, eds. 2005. *Phylogeny and Conservation.* Cambridge University Press, Cambridge, UK.

Purvis, A., J. L. Gittleman, G. Cowlishaw, and G. M. Mace. 2000. Predicting extinction risk in declining species. *Proceedings of the Royal Society of London B, Biological Sciences* 267:1947–1952.

Purvis, A., S. Nee, and P. H. Harvey. 1995. Macroevolutionary inferences from primate phylogeny. *Proceedings of the Royal Society of London B, Biological Sciences* 260:329–333.

Pybus, O. G., and P. H. Harvey. 2000. Testing macro-evolutionary models using incomplete molecular phylogenies. *Proceedings of the Royal Society of London B, Biological Sciences* 267:2267–2272.

Pyron, R. A., and J. J. Wiens. 2013. Large-scale phylogenetic analyses reveal the causes of high tropical amphibian diversity. *Proceedings of the Royal Society of London B, Biological Sciences* 280:20131622.

Questad, E. J., and B. L. Foster. 2008. Coexistence through spatio-temporal heterogeneity and species sorting in grassland plant communities. *Ecology Letters* 11:717–726.

Rabosky, D. L. 2014. Automatic detection of key innovations, rate shifts, and diversity-dependence on phylogenetic trees. *PLoS ONE* 9:e89543.

Rabosky, D. L., and E. E. Goldberg. 2015. Model inadequacy and mistaken inferences of trait-dependent speciation. *Systematic Biology* 64:340–355.

Rabosky, D. L., and I. J. Lovette. 2008. Density-dependent diversification in North American wood warblers. *Proceedings of the Royal Society of London B, Biological Sciences* 275:2363–2371.

Rabosky, D. L., F. Santini, J. Eastman, S. A. Smith, B. Sidlauskas, J. Chang, and M. E. Alfaro. 2013. Rates of speciation and morphological evolution are correlated across the largest vertebrate radiation. *Nature Communications* 4:1958.

Rafferty, N. E., and A. R. Ives. 2012. Pollinator effectiveness varies with experimental shifts in flowering time. *Ecology* 93:803–814.

Rafferty, N. E., and A. R. Ives. 2013. Phylogenetic trait-based analyses of ecological networks. *Ecology* 94:2321–2333.

Rao, C. R. 1982. Diversity and dissimilarity coefficients: A unified approach. *Theoretical Population Biology* 21:24–43.

Raunkiær, C. 1934. *The Life Forms of Plants and Statistical Plant Geography*. Clarendon, Oxford, UK.

Raup, D. M. 1979. Biases in the fossil record of species and genera. *Bulletin of the Carnegie Museum of Natural History* 13:85–91.

Raup, D. M., S. J. Gould, T.J.M. Schopf, and D. S. Simberloff. 1973. Stochastic models of phylogeny and the evolution of diversity. *Journal of Geology* 81:525–542.

Redding, D. W., K. Hartmann, A. Mimoto, D. Bokal, M. DeVos, and A. Ø. Mooers. 2008. Evolutionarily distinct species often capture more phylogenetic diversity than expected. *Journal of Theoretical Biology* 251:606–615.

Redding, D. W., and A. Ø. Mooers. 2006. Incorporating evolutionary measures into conservation prioritization. *Conservation Biology* 20:1670–1678.

Rejmanek, M., and D. M. Richardson. 1996. What attributes make some plant species more invasive? *Ecology* 77:1655–1661.

Resetarits, W. J., and J. Bernardo. 1998. *Experimental Ecology: Issues and Perspectives*. Oxford University Press, Oxford, UK.

Revell, L. J., D. L. Mahler, P. R. Peres-Neto, and B. D. Redelings. 2012. A new phylogenetic method for identifying exceptional phenotypic diversification. *Evolution* 66:135–146.

Rezende, E. L., P. Jordano, and J. Bascompte. 2007. Effects of phenotypic complementarity and phylogeny on the nested structure of mutualistic networks. *Oikos* 116:1919–1929.

Rezende, E. L., J. E. Lavabre, P. R. Guimaraes, P. Jordano, and J. Bascompte. 2007. Non-random coextinctions in phylogenetically structured mutualistic networks. *Nature* 448:925–926.

Ricciardi, A., and S. K. Atkinson. 2004. Distinctiveness magnifies the impact of biological invaders in aquatic ecosystems. *Ecology Letters* 7:781–784.

Ricciardi, A., and M. Mottiar. 2006. Does Darwin's naturalization hypothesis explain fish invasions? *Biological Invasions* 8:1403–1407.

Ricklefs, R. E. 1979. *Ecology*. Chiron, New York, NY.

Ricklefs, R. E. 1996. Phylogeny and ecology. *Trends in Ecology & Evolution* 11:229–230.

Ricklefs, R. E. 2007. History and diversity: Explorations at the intersection of ecology and evolution. *American Naturalist* 170:S56–S70.

Ricklefs, R. E., and D. Schluter, eds. 1993. *Species Diversity in Ecological Communities: Historical and Geographical Perspectives*. University of Chicago Press, Chicago, IL, US.

Ricklefs, R. E., and J. M. Starck. 1996. Applications of phylogenetically independent contrasts: A mixed progress report. *Oikos* 77:167–172.

Rodrigues, A.S.L., T. M. Brooks, and K. J. Gaston. 2005. Integrating phylogenetic diversity in the selection of priority areas for conservation: Does it make a difference? in *Phylogeny and Conservation*, ed. A. Purvis, J. L. Gittleman, and T. M. Brooks, eds, pp. 101–119. Cambridge University Press, Cambridge, UK.

Rodrigues, A.S.L., R. Grenyer, J. E. Baillie, O.R.P. Bininda-Emonds, J. L. Gittlemann, M. Hoffmann, K. Safi, et al. 2011. Complete, accurate, mammalian phylogenies aid conservation planning, but not much. *Philosophical Transactions of the Royal Society of London B, Biological Sciences* 366:2652–2660.

Rolland, J., M. W. Cadotte, J. Davies, V. Devictor, S. Lavergne, N. Mouquet, S. Pavoine, et al. 2011. Using phylogenies in conservation: New perspectives. *Biology Letters* 8:692–694.

Ronquist, F., and J. P. Huelsenbeck. 2003. MRBAYES 3: Bayesian phylogenetic inference under mixed models. *Bioinformatics* 19:1572–1574.

Roquet, C., W. Thuiller, and S. Lavergne. 2013. Building megaphylogenies for macroecology: Taking up the challenge. *Ecography* 36:13–26.

Rosauer, D., S. W. Laffan, M. D. Crisp, S. C. Donnellan, and L. G. Cool. 2009. Phylogenetic endemism: A new approach for identifying geographical concentrations of evolutionary history. *Molecular Ecology* 18:4061–4072.

Rosauer, D., and A. Mooers. 2013. Nurturing the use of evolutionary diversity in nature conservation. *Trends in Ecology & Evolution* 28:322–323.

Rosauer, D. F., S. Ferrier, K. J. Williams, G. Manion, J. S. Keogh, and S. W. Laffan. 2014. Phylogenetic generalised dissimilarity modelling: A new approach to analysing and predicting spatial turnover in the phylogenetic composition of communities. *Ecography* 37:21–32.

Ross, H. H. 1957. Principles of natural coexistence indicated by leafhopper populations. *Evolution* 11:113–129.

Russell, G. J., T. M. Brooks, M. M. McKinney, and C. G. Anderson. 1998. Present and future taxonomic selectivity in bird and mammal extinctions. *Conservation Biology* 12:1365–1376.

Sakai, K. 1965. Contributions to the problem of species colonization from the viewpoint of competition and migration, in *The genetics of colonizing species*, ed. H. G. Baker and G. L. Stebbins. Academic Press, New York, NY.

Sala, O. E., F. S. Chapin, J. J. Armesto, E. Berlow, J. Bloomfield, R. Dirzo, E. Huber-Sanwald, et al. 2000. Biodiversity: Global biodiversity scenarios for the year 2100. *Science* 287:1770–1774.

Sanderson, M. J. 2002. Estimating absolute rates of molecular evolution and divergence times: A penalized likelihood approach. *Molecular Biology and Evolution* 19:101–109.

Sanderson, M. J., M. J. Donoghue, W. Piel, and T. Eriksson. 1994. TreeBASE: A prototype database of phylogenetic analyses and an interactive tool for browsing the phylogeny of life. *American Journal of Botany* 81:183.

Sarkar, S., R. L. Pressey, D. P. Faith, C. R. Margules, T. Fuller, D. M. Stoms, A. Moffett, et al. 2006.Biodiversity conservation planning tools: Present status and challenges for the future. *Annual Review of Environmental Resources* 31:123–159.

Sax, D. F., J. J. Stachowicz, J. H. Brown, J. F. Bruno, M. N. Dawson, S. D. Gaines, R. K. Grosberg, et al. 2007. Ecological and evolutionary insights from species invasions. *Trends in Ecology & Evolution* 22:465–471.

Schamp, B. S., J. Chau, and L. W. Aarssen. 2008. Dispersion of traits related to competitive ability in an old-field plant community. *Journal of Ecology* 96:204–212.

Scheffer, M., and E. H. van Nes. 2006. Self-organized similarity, the evolutionary emergence of groups of similar species. *Proceedings of the National Academy of Science of the United States of America* 103:6230–6235.

Scheiner, S. M. 2012. A metric of biodiversity that integrates abundance, phylogeny, and function. *Oikos* 121:1191–1202.

Schluter, D. 1994. Experimental evidence that competition promotes divergence in adaptive radiation. *Science* 266:798–801.

Schluter, D. 2000. Ecological character displacement in adaptive radiations. *American Naturalist* 156:S4–S16.

Schoener, T. W. 1968. Anolis lizards of Bimini: Resource partitioning in a complex fauna. *Ecology* 49:704–726.

Schwartz, M. D. 1990. Detecting the onset of spring: A possible application of phenological models. *Climate Research* 1:23–29.

Schweiger, O., S. Klotz, W. Durka, and I. Kuhn. 2008. A comparative test of phylogenetic diversity indices. *Oecologia* 157:485–495.

Seehausen, O. 2006. African cichlid fish: A model system in adaptive radiation research. *Proceedings of the Royal Society of London B, Biological Sciences* 273:1987–1998.

Simberloff, D. S. 1970. Taxonomic diversity of island biotas. *Evolution* 24:23–47.

Simberloff, D. 1978. Using island biogeographic distributions to determine if colonization is stochastic. *American Naturalist* 112:713–726.

Simberloff, D., and W. Boecklen. 1981. Santa Rosalia reconsidered: Size ratios and competition. *Evolution* 35:1206–1228

Simms, D. A. 1979. North American weasels: Resource utilization and distribution. *Canadian Journal of Zoology* 57:504–520.

Simpson, G. G. 1944. *Tempo and Mode in Evolution*. Columbia University Press, New York, NY.

Slingsby, J. A., and G. A. Verboom. 2006. Phylogenetic relatedness limits co-occurrence at fine spatial scales: Evidence from the schoenoid sedges (Cyperaceae : Schoeneae) of the Cape Floristic Region, South Africa. *American Naturalist* 168:14–27.

Smith, F. A., A. G. Boyer, J. H. Brown, D. P. Costa, T. Dayan, S. M. Ernest, A. R. Evans, et al. 2010. The evolution of maximum body size of terrestrial mammals. *Science* 330:1216–1219.

Smith, K. G. 2006. Patterns of nonindigenous herpetofaunal richness and biotic homogenization among Florida counties. *Biological Conservation* 127:327–335.

Sodhi, N. S., D. Bickford, A. C. Diesmos, T. M. Lee, L. P. Koh, B. W. Brook, C. H. Sekercioglu, and C. J. Bradshaw. 2008. Measuring the meltdown: Drivers of global amphibian extinction and decline. *PLoS ONE* 3:e1636.

Solow, A. R., and S. Polasky. 1994. Measuring biological diversity. *Environmental and Ecological Statistics* 1:95–107.

Sørensen, T. 1948. A method of establishing groups of equal amplitude in plant sociology based on similarity of species and its application to analyses of the vegetation on Danish commons. *Biologiske Skrifter* 5:1–34.

Sorte, C.J.B., I. Ibáñez, D. M. Blumenthal, N. A. Molinari, L. P. Miller, E. D. Grosholz, J. M. Diez, et al. 2012. Poised to prosper? A cross-system comparison of climate change effects on native and non-native species performance. *Ecology Letters* 16:261–271.

Spalding, V. M. 1909. *Distribution and Movements of Desert Plants*. Carnegie Institute of Washington, Washington, DC.

Srivastava, D. S. 1999. Using local-regional richness plots to test for species saturation: Pitfalls and potentials. *Journal of Animal Ecology* 68:1–16.

Srivastava, D. S., M. W. Cadotte, A. A. M. MacDonald, N. Mirotchnick, and R. G. Marushia. 2012. Phylogenetic diversity and the functioning of ecosystems. *Ecology Letters* 15:637–648.

Stamatakis, A. 2014. RAxML version 8: A tool for phylogenetic analysis and post-analysis of large phylogenies. *Bioinformatics* 30:1312–1313.

Stebbins, G. L. 1974. *Flowering Plants: Evolution above the Species Level*. A. Edwards Arnold, London.

Stoltzfus, A., B. O'meara, J. Whitacre, R. Mounce, E. L. Gillespie, S. Kumar, D. F. Rosauer, and R. A. Vos. 2012. Sharing and re-use of phylogenetic trees (and associated data) to facilitate synthesis. *BMC Research Notes* 5:574.

Strauss, S. Y., C. O. Webb, and N. Salamin. 2006. Exotic taxa less related to native species are more invasive. *Proceedings of the National Academy of Sciences of the United States of America* 103:5841–5845.

Strecker, A. L., and J. D. Olden. 2014. Fish species introductions provide novel insights into the patterns and drivers of phylogenetic structure in freshwaters. *Proceedings of the Royal Society of London B, Biological Sciences* 281:20133003.

Streicker, D. G., A. S. Turmelle, M. J. Vonhof, I. V. Kuzmin, G. F. McCracken, and C. E. Rupprecht. 2010. Host phylogeny constrains cross-species emergence and establishment of rabies virus in bats. *Science* 329:676–679.

Strong, D. R., Jr., 1980. Null hypotheses in ecology. *Synthese* 43:271–285.

Strong, D. R., Jr., L. A. Szyska, and D. S. Simberloff. 1979. Test of community-wide character displacement against null hypotheses. *Evolution* 33:897–913.

Swenson, N. G. 2011. Phylogenetic beta diversity metrics, trait evolution and inferring the functional beta diversity of communities. *PLoS ONE* 6:e21264.

Swenson, N. G. 2014. *Functional and Phylogenetic Ecology in R.* Springer, New York, NY.

Swenson, N. G., B. J. Enquist, J. Thompson, and J. K. Zimmerman. 2007. The influence of spatial and size scales on phylogenetic relatedness in tropical forest communities. *Ecology* 88:1770–1780.

Swofford, D. L. 2003. *PAUP* 4.0b10: Phylogenetic Analysis Using Parsimony (*and Other Methods).* Sinauer Associates, MA, US.

Taper, M. L., and T. J. Case. 1992. Models of character displacement and the theoretical robustness of taxon cycles. *Evolution* 46:317–333.

Thomas, C. D., A. Cameron, R. E. Green, M. Bakkenes, L. J. Beaumont, Y. C. Collingham, et al. 2004. Extinction risk from climate change. *Nature* 427:145–148.

Thuiller, W., L. Gallien, I. Boulangeat, F. De Bello, T. Munkemuller, C. Roquet, and S. Lavergne. 2010. Resolving Darwin's naturalization conundrum: A quest for evidence. *Diversity & Distributions* 16:461–475.

Thuiller, W., S. Lavergne, C. Roquet, I. Boulangeat, B. Lafourcade, and M. B. Araujo. 2011. Consequences of climate change on the tree of life in Europe. *Nature* 470:531–534.

Tilman, D. 1982. *Resource Competition and Community Structure.* Princeton University Press, Princeton, NJ.

Tilman, D. 1999. The ecological consequences of changes in biodiversity: A search for general principles. *Ecology* 80:1455–1474.

Tilman, D. 2004. Niche tradeoffs, neutrality, and community structure: A stochastic theory of resource competition, invasion, and community assembly. *Proceedings of the National Academy of Science of the United States of America* 101:10854–10861.

Tilman, D., D. Wedin, and J. Knops. 1996. Productivity and sustainability influenced by biodiversity in grassland ecosystems. *Nature* 379:718–720.

Tucker, C. M., and M. W. Cadotte. 2013. Unifying measures of biodiversity: Understanding when richness and phylogenetic diversity should be congruent. *Diversity and Distributions* 19:845–854.

Tucker, C. M., M. W. Cadotte, S. B. Carvalho, T. J. Davies, S. Ferrier, S. A. Fritz, R. Grenyer, et al. 2016. A guide to phylogenetic metrics for conservation, community ecology and macroecology. *Biological Reviews* (in press).

Tucker, C. M., M. W. Cadotte, T. J. Davies, and A. G. Rebelo. 2012. The distribution of biodiversity: Linking richness to geographical and evolutionary rarity in a biodiversity hotspot. *Conservation Biology* 26:593–601.

Tuomisto, H. 2010. A diversity of beta diversities: Straightening up a concept gone awry. Part 1. Defining beta diversity as a function of alpha and gamma diversity. *Ecography* 33:2–22.

Turelli, M., N. H. Barton, and J. A. Coyne. 2001. Theory and speciation. *Trends in Ecology & Evolution* 16:330–343.

Udvardy, M.F.D. 1959. Notes on the ecological concepts of habitat, biotope and niche. *Ecology* 40:725–728.

Underwood, A. J. 1997. *Experiments in Ecology*. Cambridge University Press, Cambridge, UK.

Valente, L. M., G. Reeves, J. Schnitzler, I. P. Mason, M. F. Fay, T. G. Rebelo, M. W. Chase, and T. G. Barraclough. 2009. Diversification of the African genus *Protea* (Proteaceae) in the Cape biodiversity hotspot and beyond: Equal rates in different biomes. *Evolution* 64:745–760.

Vamosi, J. C., and J.R.U. Wilson. 2008. Nonrandom extinction leads to elevated loss of angiosperm evolutionary history. *Ecology Letters* 11:1047–1053.

Vamosi, S. M., S. B. Heard, J. C. Vamosi, and C. O. Webb. 2009. Emerging patterns in the comparative analysis of phylogenetic community structure. *Molecular Ecology* 18:572–592.

Van Noorden, R., B. Maher, and R. Nuzzo. 2014. The top 100 papers. *Nature* 514:550–553.

Van Valen, L. 1971. Adaptive zones and the orders of mammals. *Evolution* 25:420–428.

Van Valkenburgh, B. 1999. Major patterns in the history of carnivorous mammals. *Annual Review of Earth and Planetary Sciences* 27:463–493.

Van Valkenburgh, B., and R. K. Wayne. 1994. Shape divergence associated with size convergence in sympatric East-African jackals. *Ecology* 75:1567–1581.

Vane-Wright, R. I., C. J. Humphries, and P. H. Williams. 1991. What to Protect? Systematics and the agony of choice. *Biological Conservation* 55:235–254.

Veech, J. A. 2012. Significance testing in ecological null models. *Theoretical Ecology* 5:611–616.

Veech, J. A., K. S. Summerville, T. O. Crist, and J. C. Gering. 2002. The additive partitioning of species diversity: Recent revival of an old idea. *Oikos* 99:3–9.

Via, S. 2001. Sympatric speciation in animals: The ugly duckling grows up. *Trends in Ecology & Evolution* 16:381–390.

Violle, C., D. R. Nemergut, Z. Pu, and L. Jiang. 2011. Phylogenetic limiting similarity and competitive exclusion. *Ecology Letters* 14:782–787.

Vitousek, P. M., H. A. Mooney, J. Lubchenco, and J. M. Melillo. 1997. Human domination of Earth's ecosystems. *Science* 277:494–499.

Von Euler, F. 2001. Selective extinction and rapid loss of evolutionary history in the bird fauna. *Proceedings of the Royal Society of London B, Biological Sciences* 268:127–130.

Walker, B., A. Kinzig, and J. Langridge. 1999. Plant attribute diversity, resilience, and ecosystem function: The nature and significance of dominant and minor species. *Ecosystems* 2:95–113.

Wang, X., N. G. Swenson, T. Wiegand, A. Wolf, R. Howe, F. Lin, J. Ye, et al. 2013. Phylogenetic and functional diversity area relationships in two temperate forests. *Ecography* 36:883–893.

Warming, E. 1909. *Oecology of Plants: An Introduction to the Study of Plant Communities*. Clarendon, Oxford, UK.

Warwick, R. M., and K. R. Clarke. 1998. Taxonomic distinctness and environmental assessment. *Journal of Applied Ecology* 35:532–543.

Watt, A. S. 1947. Pattern and process in the plant community. *Journal of Ecology* 35:1–22.

Webb, C. O. 2000. Exploring the phylogenetic structure of ecological communities: An example for rain forest trees. *American Naturalist* 156:145–155.

Webb, C. O., D. D. Ackerly, M. A. McPeek, and M. J. Donoghue. 2002. Phylogenies and community ecology. *Annual Review of Ecology and Systematics* 33:475–505.

Webb, C. O., and M. J. Donoghue. 2005. Phylomatic: Tree assembly for applied phylogenetics. *Molecular Ecology Notes* 5:181–183.

Webb, C. O., J. B. Losos, and A. A. Agrawal. 2006. Integrating phylogenies into community ecology. *Ecology* 87:S1–S2.

Weiher, E., and P. A. Keddy. 1995a. The assembly of experimental wetland plant-communities. *Oikos* 73:323–335.

Weiher, E., and P. A. Keddy. 1995b. Assembly rules, null models, and trait dispersion: New questions from old patterns. *Oikos* 74:159–164.

Weir, J. T., and D. Schluter. 2007. The latitudinal gradient in recent speciation and extinction rates of birds and mammals. *Science* 315:1574–1576.

Weitzman, M. L. 1998. The Noah's Ark problem. *Econometrica* 66:1279–1298.

Westoby, M., M. R. Leishman, and J. M. Lord. 1995. On misinterpreting the phylogenetic correction. *Journal of Ecology* 83:531–534.

Whittaker, R. H. 1960. Vegetation of the Siskiyou Mountains, Oregon and California. *Ecological Monographs* 30:279–338.

Whittaker, R. J., and K. A. Triantis. 2012. The species–area relationship: An exploration of that "most general, yet protean pattern." *Journal of Biogeography* 39:623–626.

Wiens, J. J., D. D. Ackerly, A. P. Allen, B. L. Anacker, L. B. Buckley, H. V. Cornell, E. I. Damschen, et al. 2010. Niche conservatism as an emerging principle in ecology and conservation biology. *Ecology Letters* 13:1310–1324.

Wiens, J. J., and M. J. Donoghue. 2004. Historical biogeography, ecology, and species richness. *Trends in Ecology & Evolution* 19:639–644.

Wiens, J. J., and C. H. Graham. 2005. Niche conservatism: Integrating evolution, ecology, and conservation biology. *Annual Review of Ecology Evolution and Systematics* 36:519–539.

Wiley, E. O., and J. T. Collins. 1991. *The Compleat Cladist: A Primer of Phylogenetic Procedures.* Museum of Natural History, University of Kansas, Lawrence, KS, US.

Williams, C. B. 1947. The generic relations of species in small ecological communities. *Journal of Animal Ecology* 16:11–18.

Willig, M. R., D. M. Kaufman, and R. D. Stevens. 2003. Latitudinal gradients of biodiversity: Pattern, process, scale, and synthesis. *Annual Review of Ecology Evolution and Systematics* 34:273–309.

Willmer, P. 2012. Ecology: Pollinator-plant synchrony tested by climate change. *Current Biology* 22:R131–R132.

Wilson, E. O. 1965. The challenge from related species, in *The Genetics of Colonizing Species*, ed. H. G. Baker and G. L. Stebbins, pp. 7–27. Academic Press, New York, NY.

Wilson, E. O. 1992. *The Diversity of Life.* Belknap, Cambridge, MA, US.

Wilson, J. B. 1987. Methods for detecting non-randomness in species co-occurrences: A contribution. *Oecologia* 73:579–582.

Winter, M., V. Devictor, and O. Schweiger. 2013. Phylogenetic diversity and nature conservation: Where are we? *Trends in Ecology & Evolution* 28:199–204.

Winter, M., O. Schweiger, S. Klotz, W. Nentwig, P. Andriopoulos, M. Arianoutsou, C. Basnou, et al. 2009. Plant extinctions and introductions lead to phylogenetic and taxonomic homogenization of the European flora. *Proceedings of the National Academy of Sciences of the United States of America* 106:21721–21725.

Witting, L., and V. Loeschcke 1995. The optimization of biodiversity conservation. *Biological Conservation* 71:205–207.

Wolkovich, E. M., B. I. Cook, J. M. Allen, T. M. Crimmins, J. L. Betancourt, S. E. Travers, S. Pau, et al. 2012. Warming experiments underpredict plant phenological responses to climate change. *Nature* 485:494–497.

Woodburne, M. O., T. H. Rich, and M. S. Springer. 2003. The evolution of tribosphy and the antiquity of mammalian clades. *Molecular Phylogenetics and Evolution* 28:360–385.

Wright, J. P., S. Naeem, A. Hector, C. Lehman, P. B. Reich, B. Schmid, and D. Tilman. 2006. Conventional functional classification schemes underestimate the relationship with ecosystem functioning. *Ecology Letters* 9:111–120.

Yang, Z. 2014. *Molecular Evolution: A Statistical Approach.* Oxford University Press, Oxford, UK.

Yessoufou, K., T. J. Davies, O. Maurin, M. Kuzmina, H. Schaefer, M. Bank, and V. Savolainen. 2013. Large herbivores favour species diversity but have mixed impacts on phylogenetic community structure in an African savanna ecosystem. *Journal of Ecology* 101:614–625.

Yom-Tov, Y. 1991. Character displacement in the psammophile Gerbillidae of Israel. *Oikos* 60:173–179.

Zachos, F. E., and J. C. Habel, eds. 2011. *Biodiversity Hotspots: Distribution and Protection of Conservation Priority Areas.* Springer, Heidelberg, Germany.

Zavaleta, E. S., and K. B. Hulvey. 2007. Realistic variation in species composition affects grassland production, resource use and invasion resistance. *Plant Ecology* 188:39–51.

Zavaleta, E. S., J. R. Pasari, K. B. Hulvey, and G. D. Tilman. 2010. Sustaining multiple ecosystem functions in grassland communities requires higher biodiversity. *Proceedings of the National Academy of Sciences of the United States of America* 107:1443–1446.

Zoltan, B. D. 2005. Rao's quadratic entropy as a measure of functional diversity based on multiple traits. *Journal of Vegetation Science* 16:533–540.

INDEX